Placing Animals

HUMAN GEOGRAPHY IN THE TWENTY-FIRST CENTURY ISSUES AND APPLICATIONS

SERIES EDITOR
Barney Warf, University of Kansas

Human geography is increasingly focused on real-world problems. Applying geographic concepts to current global concerns, this series focuses on the urgent issues confronting us as we move into the new century. Designed for university-level geography and related multidisciplinary courses such as area studies, global issues, and development, these textbooks are richly illustrated and include suggestions for linking to related Internet resources. The series aims to help students to better understand, integrate, and apply common themes and linkages in the social and physical sciences and in the humanities, and, by doing so, to become more effective problem solvers in the challenging world they will face.

Placing Animals

An Introduction to the Geography of Human-Animal Relations

Julie Urbanik

ROWMAN & LITTLEFIELD PUBLISHERS, INC.
Lanham • Boulder • New York • Toronto • Plymouth, UK

Published by Rowman & Littlefield Publishers, Inc.
A wholly owned subsidiary of The Rowman & Littlefield Publishing Group, Inc.
4501 Forbes Boulevard, Suite 200, Lanham, Maryland 20706
www.rowman.com

10 Thornbury Road, Plymouth PL6 7PP, United Kingdom

British Library Cataloguing in Publication Information Available

Library of Congress Cataloging-in-Publication Data Available

ISBN 978-1-4422-1184-1 (cloth : alk. paper)
ISBN 978-1-4422-1185-8 (pbk. : alk. paper)
ISBN 978-1-4422-1186-5 (electronic)

∞™ The paper used in this publication meets the minimum requirements of American
National Standard for Information Sciences—Permanence of Paper for Printed Library
Materials, ANSI/NISO Z39.48-1992.

Printed in the United States of America

For Rooster and Lewis,
there have been, are, and will be others,
but they will never be you.

Contents

Acknowledgments

This book would not exist save for the support shown to me by multiple humans and animals. I would like to thank Barney Warf, the editor of this series, who encouraged me to propose a book on animal geography in the first place. Susan McEachern, the book's editor, and her assistant, Grace Baumgartner, have provided excellent support and feedback throughout the process.

Thanks are also due to the students at UMKC who put up with me over and over again as I tried to figure out my way through the material. I couldn't have found a more enthusiastic, intelligent, funny, patient, and thoughtful group of students anywhere. Nazgol Bagheri, Jesse Nikkel, and Trina Weilert are GIS heroes. Mary Ann Denzer, Trinity Koger, Mary Morgan, Aaron Phillips, Danielle Schreck, Rebecca Varady, and Alex Waechter are research and sounding board heroes. They all learned more about human-animal relations than they probably thought was possible, but their research skills, professionalism, and attention to detail helped make my job infinitely easier. I am also grateful to UMKC and the Department of Geosciences who supported my efforts with much needed time.

The love and support from my parents, sister, brother, stepchildren, and mother-in-law have been nonstop from day one and I am deeply thankful for their encouragement. I do not have the words to express how profound my gratitude is to my husband for all that he has done for me during this process. All I can say is that it makes for a funny story now that it's over. Finally, Moonshine, Roxy, and Ziggy have kept me grounded in the animal world. Their feline advice to take lots of baths, stretch, play, swat something occasionally, and meditate regularly—preferably on one's back in a sunny spot—was spot on.

Preface

Animals surround me right now as I write these words: Inside are three cats; sculptures of elephants, cats, water buffalo, frogs, birds, and an octopus; photos of cheetahs, elephants, seals, giraffes, and all sorts of birds; and a painting of coyotes. Pieces of animals decorate nearly every room (all found!)—bird nests, a porcupine quill, bison fur, a wild-turkey eggshell, too many feathers, a chip from a tree that had been visited by a beaver, seashells, pieces of turtle shell, a jaguar whisker, and the skeletal mouth of a sea urchin. Outside there are butterflies, a huge spider that lives by the porch light, mosquitoes, blue jays, cardinals, three species of woodpeckers, three species of finches, nuthatches, worms, crickets and other creepy-crawlies and creepy-fliers, starlings, hummingbirds, chipmunks, squirrels, and occasionally our resident opossum, a Cooper's hawk, and the neighborhood bully cats. Furthermore, there is milk and cheese in the refrigerator, cat food made of cows, chickens, turkeys, salmon, and tuna, honey, leather shoes, a leather softball glove, and household products that have been tested on animals.

I suspect that your list might also be similar—give or take a few found animal parts. Your list might include horses and snakes, or fishing gear and guns, or perhaps a fur coat or a freezer full of chicken or a trash can full of hamburger wrappers. No matter—the animals are there. This book is an invitation to see and reflect on your own particular relationships with nonhumans and why you have them even as it is an invitation to see how human societies relate to the nonhuman world. Moreover, this book is an invitation to consider animals from a geographic perspective. Geography has a long history of studying animals with its emphasis on understanding the physical and human components that make up our planet, but somewhat surprisingly this book is the first to focus exclusively on the spectrum of human-animal geographies that are out there. While it is written primarily for undergraduate and beginning graduate students, I hope that it will find its way to anyone with an interest in learning about how humans relate to the animal species around them. I have only just begun learning how to teach about animal geographies, but what I have learned thus far is that most people have never had a chance to think critically or intellectually about the animals around them. Obviously they are not incapable of doing so, but very few people even know all the places to look. Hence geography is so fundamental to studying humans and animals because we must learn where and how to look and what (or whom) to look for before we can credibly reflect on what we like, agree with, or reject from what we find. This book, then, is an invitation to join an expedition to explore and map the known and unknown terrain that is the human-animal landscape.

Chapter One

Geography and
Human-Animal Relations

American naturalist Henry Beston once wrote that animals "are not brethren, they are not underlings; they are other nations, caught with ourselves in the net of life and time, fellow prisoners of the splendor and travail of the earth" (1992, 25). Who are these "other nations" that we, as human animals, share our world with? Opossum. Tarantula. Lynx. Macaw. Elephant. Shrew. Coyote. Anaconda. Penguin. Chimpanzee. Each one of these species calls up a constellation of images, experiences, phrases, and emotions—simultaneously our own and the result of the cultures we live in. We eat them, wear them, live with them, work them, experiment on them, try to save them, spoil them, abuse them, fight them, hunt them, buy, sell, and trade them, and love, fear, or hate them. Where, how, and why do we have the relationships that we do with different animals? Why are some animals food and some animals pets? Why are some animals both? Do we have obligations to other species? Do some animals matter more than others? This book uses the lens of animal geography to think through these questions and to promote a critical understanding of the role that other animals play in our human lives.

What are some of the cultural legacies that have shaped the interactions we have with other species today? Consider for a moment that the earliest known cave art contains images of animals but no humans (shamans or spirits yes, but no humans). Around thirty-one thousand years ago the Aurignacian culture in southern France painted stunning images of animals such as horses, bears, and rhinoceroses along the contours of the Chauvet cave wall (Clottes and Féruglio 2011). Between five thousand and ten thousand years ago, the San people painted detailed portraits of giraffes, elands, kudus, ostriches, and more at the Cave of Inanke in what is today Matobo National Park in southern Zimbabwe (FitzGerald 2010). In ancient Egypt (three thousand to five thousand years ago), tombs, temples, and stele were heavily decorated with a pantheon of human-animal gods such as Anubis, God of the dead with a jackal head and human body, and Thoth, God of wisdom and writing with an ibis head and human body. Even the zodiac (Greek for "animal-figure") systems of the Greeks and Chinese and the Mayan calendar made use of animals in helping these cultures navigate time, social practices, and belief systems.

Individuals have taken different views throughout history—views that also shape our practices today. Pythagoras, the ancient Greek intellectual who lived during the sixth century BCE was an ardent vegetarian. In fact, vegetarians were called Pythagoreans at the time. A few centuries later the Romans, under Titus, would inaugurate the opening of the Colosseum in 81 CE with a hundred-day spectacle that included killing nine thousand wild animals (Muller 2011). In Europe during the 1500s two French intellectuals presented opposing views of animals. René Descartes saw them as automatons because they could not speak human languages and therefore he concluded they could not reason. He believed a difference in kind existed between humans and animals. Michel Montaigne, however, found the difference to be one of degree, not kind, and recognized—especially by watching his cats—that animals appear to have a level of "self" that makes them much more than mechanical entities like clocks (which is what Descartes compared dogs to). A few centuries later, British naturalist Charles Darwin brought human-animal relationships to the fore in a new way with his 1859 publication *On the Origin of Species*, which outlined his theory of evolution through natural selection. This theory placed humans not above, or even separate from, animals, but as simply one result of evolutionary processes. Shortly thereafter, in 1887, Anna Sewell's book *Black Beauty* was published, becoming the first known piece of literature written from the perspective of an animal—in this case a horse documenting its treatment by loving and abusive owners.

In our time, visibly or invisibly, animals swirl around our everyday lives in myriad ways. The international television channel Animal Planet perhaps most broadly encompasses individual and cultural human-animal relationships. The channel is dedicated to all things animal and with its stories of sea life covered in oil, polar bears struggling on smaller and smaller pieces of ice, football players involved in dogfighting rings, pets who saved their owners, and hilarious or terrifying animal antics, it exemplifies our convoluted relations with other species. These convoluted relations are undergirded by three key points about human-animal relations that are as true now as they have been historically. First, the boundary between humans and animals is not consistent. We, as humans, are biologically animals yet most societies on the planet today separate humans from nonhuman animals in strange and oftentimes contradictory ways by blurring the similarities or spotlighting the differences. For example, popular culture like Disney creates animal characters that act like humans even though giving real nonhumans human characteristics, or anthropomorphizing them, is often frowned upon. Animals that are used for biomedical research are seen as enough like humans to be experimented upon yet different enough from humans that they are outside of research constructs that claim it is unethical to experiment upon humans. These boundaries differ greatly from cultures where animals can be gods. Second, animals are much more than simply background to human lives only to be acknowledged intermittently. They are, in fact, quite central to most people's everyday existence. Where would we be without bees to pollinate our agricultural crops, or without animal skins to wrap our feet or bodies in? We evolved into modern humans partly because of our ability to adapt and hunt other species as well as domesticate them. Third, who and where you are as a human in the world shapes the type of interaction you will have with different species. For example, in some parts of the world dogs are adored as pets, while in others they are raised in cages like chickens or pigs for the sole purpose of human consumption. What is the difference between eating a dog or a pig? They are both highly intelligent beings who have been domesticated for thousands of years, but which one (or both) you will eat depends on your geography. Geography, in fact, undergirds all three of these points.

This book is premised on the fact that geography is central both to our everyday interactions with animals and to academic interest in understanding the variety of human-animal relationships around the world. As an academic discipline, geography has historically focused on the relationships between the natural world and human society, yet human-animal relationships have, until recently, been narrowly constructed as biological studies of species distributions or as utilitarian relations between human groups and livestock. The rise of the "new" animal geography over the past two decades encompasses these aspects while dramatically expanding the areas of focus by recognizing that human-animal relations are simultaneously biological, cultural, economic, ethical, geographical, and political. The purpose of this book, then, is to introduce the broad subfield of animal geography so you gain a thorough understanding of (1) the relationship between animal geography and the larger animal studies academic community, (2) the myriad geographies of human-animal interactions around the world, (3) the way in which animal geography is both challenging and contributing to the major fields of human and nature-society geography, and (4) the role of place in shaping human-animal interactions. By this I mean the full geographic conception of place, which includes both material (e.g., zoo, slaughterhouse, home, wild) and symbolic (e.g., theoretical, scientific, literary, moral) locations. Whether we are discussing the ways in which the conceptual placement of certain animals as food links to the structure of animal agriculture or how biologists' studies of properly placed habitat boundaries enables or constrains human-wildlife conflicts, how we treat nonhuman others is fundamentally rooted in the places in which we can, or cannot, interact. The remainder of this chapter will lay the groundwork for this exploration of animal geography by providing an overview of geography and animal studies. We will begin with considering the need for animal geography at all.

WHY ANIMAL GEOGRAPHY IS RELEVANT TODAY

Prominent animal geographers Jennifer Wolch and Jody Emel (1998) argue that three main reasons can be given as to why animals and human-animal relations are an increasingly visible topic in geography today. The first reason has to do with ever more sophisticated scientific understandings of how humans, through their economic production and consumption patterns, are contributing to environmental problems, and thereby our impact on other animal species specifically is becoming more pronounced. For example, a recent United Nations (UN) report, *Livestock's Long Shadow* (Steinfeld et al. 2006), analyzes the ways in which both industrial and pastoral livestock-keeping practices are contributing to human-induced climate change. There are approximately fifty billion livestock animals on the planet in any given year and the methane gas belched from cattle alone impacts our atmosphere in a negative way because it is such a potent greenhouse gas. In addition, rain forests are still being cleared to produce livestock feed. This clear cutting reduces the amount of trees available to absorb carbon dioxide as well as releasing carbon dioxide from decaying stumps and burning brush. Intensive herding practices are contributing to desertification in places like the Sahel (just below the Saharan desert). While this UN report is by no means complete in its analysis, it is the first official attempt from an international political entity to assess the environmental consequences of fifty billion livestock animals competing with humans and other living beings for resources. The global trade in wildlife—whether legal or illegal, live or dead—is also having a problematic impact. Literally millions of animals are extracted from terrestrial and marine environments every year mostly for the international pet trade, the exotic meat/parts markets, and medicinal uses. This level of extraction has environmental impacts as it changes local ecosystems, reduces local biodiversity, and decreases local people's ability to harvest the animals right around them.

A second reason for the increased scrutiny of the human-animal relationship comes from a different part of the academy—the social rather than the hard sciences. In both social sciences and humanities, changes in our understanding of how human society works have been taking place over the last several decades. Advances in social theories—mapping how social forces work as a whole—have led to a critique of what has come to be called the modernist view of the world. This view, rooted in the European Enlightenment and Scientific Revolution, values human rationality over emotion and the individual over the collective. This ardent humanism takes as its starting point the separation of humans and animals and then places humans above animals in a dualistic hierarchy of value. This view also has separated human society from nature in an explicit division, putting humans "outside" of nature. While this view did, and has, helped advance humanity and our understanding of the physical world in innumerable ways, it has also come under heavy scrutiny (1) for homogenizing human experience when clearly vast differences occur among people's experiences, shaped by gender roles, racial histories, colonial histories, sexual identities, and so on, *and* (2) by denying the interconnectedness of humans and the planet. As social theories have moved into a "postmodern" period that differentiates humans' subjective experiences, so too are they moving into a "posthuman" period with challenges to notions of what has historically been seen to separate humans from animals coming almost daily from academic disciplines such as conservation biology and ethology—the study of animal behavior—and in medical genetic research. Once-key markers of human superiority and uniqueness— tool use, language, abstract thought, even culture itself as the ability to pass down behaviors—are being debunked by species as diverse as chimpanzees, elephants, and crows. From the social sciences, posthumanism (which also includes the relations between humans and technologies) has emerged as a challenge to what constitutes a subject and how we might live in a more-than-human world. As a key result of this shift, animals are now being seen in a way they have never been within the academy: they are no longer only objects to be studied and categorized, but their experiential lives now count as do our experiences with them.

The third reason has to do with the politics of animal issues. Figure 1.1 shows where in the world animal rights and animal welfare organizations are active. The difference between animal rights and animal welfare often depends on how an individual group defines itself, but in general we can understand animal rights groups to be more focused on ending a particular animal practice while welfare groups tend to be focused more on education and humane treatment. This spectrum comes down to ethical frameworks. Some people believe that using animals for any reason is wrong—that animals have "rights" on par with humans. Other people believe that using animals in a variety of ways is acceptable, but that we must be careful to treat them properly and minimize environmental harm. A third group of people believe they do

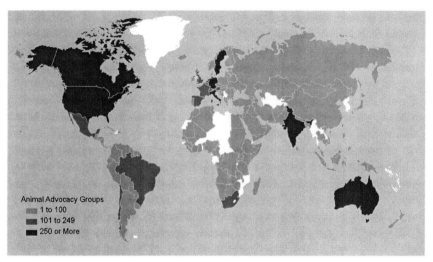

Figure 1.1.   Global Distribution of Animal Rights and Animal Welfare Groups. *Source:* **This map was compiled using the World Animal Net online directory.**

not have to be concerned about the treatment of animals at all—that humans can use other species however they see fit. As you can imagine, these differences cause a lot of political conflict. Battles over everything from animals in circuses to wildlife conservation to meat eating and fur wearing are going on all the time. While we have already seen that the conflict over the treatment and use of animals has been around for quite some time, our globally connected world with today's Internet and twenty-four-hour television has made publicizing perceived injustices easier.

Even the very definition of an animal is complicated via politics. Table 1.1 shows how varied the definitions are just in the United States. Nearly half of the states make an explicit exception for humans in defining animals, and many states go so far as to exclude whole groups of animals (e.g., livestock, fish) from being considered animals. Internationally, the laws can be just as varied with China defining an animal as "any mammal, bird, reptile, amphibian, fish or any other vertebrate or invertebrate whether wild or tame," Israel as "any vertebrae excluding man," and Zimbabwe as "any kind of domestic vertebrate animal; any kind of wild vertebrate animal in captivity" (MSU 2011). Why does getting a consistent definition of an animal seem so hard, and why do governments feel the need to explicitly exclude humans in their definitions? This is a question of politics. Many animals are excluded from legal definitions so that they can be excluded from animal welfare laws to help locally based animal-related industries like fishing or animal research. The explicit separation between humans and animals can also be based upon

Table 1.1.   Legal Definitions of "Animal" in the United States

Definition of animal	**Federal:** any live or dead dog, cat, nonhuman primate, guinea pig, hamster, rabbit, or any warm-blooded animal used for research, teaching, testing, experimentation, or exhibition purposes, or as a pet.
	States:
	1. mammal, bird, reptile, or amphibian (AR, OK, OR, SD, WA)
	2. every living/dumb/brute creature (CA, CO, FL, MD, MI, NJ, NY, RI, VT)
	3. every living being/animal (DE, HI, KY, ND, ME, MN, MS, NC, NV, TX, WI)
	4. living vertebrate (AK, AR, IA, ID, IN, KS, MO, NE, SC, UT, VA)
States that specifically exclude humans in their definitions	Arkansas, Hawaii, Idaho, Illinois, Indiana, Kansas, Maine, Maryland, Minnesota, Missouri, Montana, Nevada, New York, North Dakota, Oklahoma, Oregon, Rhode Island, South Carolina, South Dakota, Vermont, Washington, Wisconsin
Other specific exceptions	**Federal:** cold-blooded species (amphibians and reptiles), birds, rats, mice, horses not used for research purposes, farm animals, fish, and invertebrates.
	Arkansas—excludes fish
	Alaska—excludes fish
	Delaware—excludes fish, crustaceans, and mollusks
	Iowa—excludes livestock, game, furbearers, fish, reptiles, and amphibians (unless owned as pets)
	New Mexico—excludes insects and reptiles
	Texas—excludes uncaptured wild creatures
	Utah—excludes agricultural animals and wildlife
	Wisconsin—excludes reptiles and amphibians

Source: The Michigan State University Animal Legal and Historical Center online database (MSU 2011).

religious views intertwining with legal views. Regardless, the politics of defining an animal is very closely linked to resulting treatment; hence the reason that so many groups around the world are advocating for legal systems to change in one way or another.

To Wolch and Emel's (1998) three reasons we will add a fourth—the increasing acceptance of humans' emotional connections with animals. Environmental awareness, animal-based social movements, and the decentering of the human subject have all contributed to an outpouring of emotional attachment to other species. Consider the increasing role of pets, especially in industrialized and industrializing countries like the United States and China. The pet economy is worth billions in the United States alone as large segments of the population have changed from seeing common pets such as dogs and cats as protectors or mouse-catchers to something more akin to family members. Think of the extraordinary lengths people will go to save pods of beached whales, injured wildlife, abused pets, or animals harmed in a natural disaster such as an oil spill. Or what of the cultural practices that link animals and humans such as getting a tattoo of a favorite animal, collecting animal-related art, or even the profusion of wine labels with animals? Many humans find something deeply rewarding about interacting with another species, and what has often been dismissed as emotional sentimentality is becoming more and more the norm. This is not to gloss over the very contradictory ways this love is shown—what we will be exploring over the next several chapters— but highlighting the emotional connection humans have with other species is just as important as highlighting the economic, political, or intellectual relations. After all, as humans we are experiential beings and whether or not we experience terror, curiosity, awe, tenderness, empathy, hatred, friendship, indifference, or love toward other species, these experiences directly shape and are shaped by our animal encounters.

Many animal geographers today were drawn to the discipline of geography because of its focus on working to synthetically understand the way the world works: the human-animal relationship is one of the key features that have marked our history and interaction with the environment. Many animal geographers will also express a certain affinity toward animals and a pleasure in being able to combine their intellectual and personal interests, perhaps even considering themselves animal activists; but these are not prerequisites for studying the geography of human-animal relations any more than we would consider base jumping or free-climbing rock faces prerequisites for the study of geology. As we will see, the use of geography to study human-animal relations opens plenty of doors in a multitude of directions. The aim of this book is to learn how to think like a geographer when it comes to human-animal relations. Where you end up will depend upon your own location, interests, history, and identity. Let's turn now to the field of geography itself.

UNDERSTANDING GEOGRAPHY AND BUILDING A GEOGRAPHER'S TOOLKIT

Geography, from the Greek for "earth writing" or "earth description," is an academic discipline that deals with the description, distribution, and interaction of natural and human systems on the planet. In essence, geography is concerned with why things are the way they are where they are. To understand animal geography, we need to understand several points: the branches of the discipline, key geographic concepts, key analytic categories, and the major methodologies geographers use to study the world. Figure 1.2 provides a visual summary for our discussion. Before mapping all of these points, one must note that none of these operates entirely in isolation. Geographers may highlight one area for a particular reason, but often geographic research is a synthetic process in which concepts or categories overlap in different ways that reveal new perspectives on what, why, when, and where. Furthermore, the following paragraphs are merely an overview of geography. They are not intended to be an in-depth exploration of the intricacies within and between each point. Two excellent resources to build a deeper understanding of geography are the *Dictionary of Human Geography* (Gregory et al. 2009) and the *Dictionary of Physical Geography* (Thomas and Goudie 2000).

Four major branches constitute the discipline of geography today. First, the physical or natural science branch focuses on understanding the physical processes of the planet including biogeography, geology, and atmospheric sciences. Second, the human or social science and humanities component studies human cultures and how cultural practices such as religion, language, politics, art, or economic systems vary across time and space. Geography has also had the longest intellectual history of studying the relationship between humanity and the physical environment—the third branch. This branch involves understanding how the physical environment has shaped the development of human cultures, how humans are impacted by natural events such as earthquakes and floods, how humans have modified the physical world through practices such as domestication or urban development, and also how

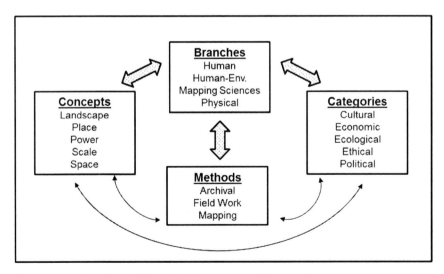

Figure 1.2.   Overview of the Main Components of Geography.

human belief systems and attitudes have shaped environmental practices. For example, if your culture believes that animals have souls like humans then you might not eat them whereas if you are in a culture that does not believe animals have souls like humans then they would be acceptable to consume. Finally, geography has been the discipline of maps and mapping sciences. Cartography, or mapmaking, has been foundational to knowledge seekers and those in power for millennia. Maps are the visual medium by which we convey our ever-changing knowledge of the wheres and whys of the world. While it might seem as if we already know where everything is today, consider the fact that southern Sudan voted in January of 2011 to secede from Sudan to create a new country. When that process is completed, the maps will have to be redrawn. Maps are constantly being revised to reflect new understandings of both physical and human systems. Newer technologies like geographic information systems (GIS) and remote sensing (using satellite data) are developing ever more sophisticated ways of visually understanding our world by enabling geographers to digitize and layer map data in novel ways.

Within these four branches, geographers use, to a varying amount, five basic concepts that help reveal the patterns of the world. You will need them all as a foundation to further learning about human-animal relations and animal geography. Scale is a concept most of us are familiar with from reading maps. On a map, scale refers to the size of the object as represented on the map in relation to the actual object itself. Scale, however, can also be understood as a relational tool of analysis moving back and forth on a continuum between the intimately local body, through the home, the neighborhood, the city, the county, the state/ province, the country, the continent, to the globe. Scale can also be used to describe amounts or levels of production of goods. Geographers often zoom in and out across scales to pinpoint what is happening at a specific level of analysis but also to seek an understanding of the relationships between scales (i.e., how local behavior influences what a country does or vice versa).

Space, another concept key to geographers, can also be understood in multiple ways. The first way is to see that space has to do with the relations between objects or events, in essence the distribution of a particular topic (whether physical or human) in the container that is the planet. This spatial analysis allows geographers to see where things are happening in the world in order to begin to investigate why. The second understanding of space that is relevant here is that space can denote locations in the abstract or the general. We can talk of zoos as a space in which a certain coming together of human and animal relations occurs, and we can even map where zoos are in the world, useful to discussions of ethics or politics perhaps, but zoo as a space is different from a specific zoo in a particular location. The Paris zoo is, instead, a place—another key term.

We can understand place as the unique and as a specific confluence of physical and human histories. The Paris zoo is a zoo space like others, yes, but it began at a specific time with a specific group of people in a specific environment that differs from the United States or Chile. The term *region* is used by geographers as a way to convey a collection of similarly organized spaces or places (e.g., the political region of North America or the region of the equatorial rain forest).

Two additional concepts are landscape and power. Landscape constitutes all the visible features of a location—the geology, flora, fauna, and the built human environment. Landscapes provide a visual record of physical and human development. We could study the landscape changes brought about by beavers building dams and creating new wetlands, or we could study the cultural animal landscape by examining where and how we see animals (or their representations) around us—on farms, in pet stores, on billboards, in dog parks, and so on. From this we can get a temporal sense of human-animal relationships in particular places. Power has to do with who or what controls spaces or places and how that power is manifested—culturally, legally, or physically. Power is intertwined with all of the analytic categories we are going to be using, and as animal geographers, we must learn to understand where and how power works across the matrix of human-animal relations. For example, humans have power over their pets in terms of providing food and shelter, but don't those same pets have a certain power over humans in terms of forcing humans into certain behaviors (exercising, playing)? Legally we construct places such as Yellowstone National Park in the United States where the government controls park activities; however, the bison and wolves that inhabit the park often exert their own form of power and leave the park's boundaries. Power, then, like all of these geographic concepts is multifaceted. As you move through the book you will continue to develop your understanding of these concepts and how they related to our next point—that of analytic categories.

Analytic categories are those umbrella categories that help geographers make sense of processes and differences around the world. This book will emphasize five major categories: cultural, ecological, economic, ethical, and political. Cultural geography explores a wide variety of learned social behaviors from religion to art to ethnic and gender identities—the full spectrum of human experience—to understand how these practices evolve, expand, contract, and exist through time and space. From this perspective we might consider the different cultures of pet keeping over time and around the world. Which cultures have pets? Why don't others? When did pet ownership become the norm? Ecological analysis has to do with understanding how environmental processes work from a natural science perspective. Here, the study of human-animal relations often requires understanding the biological and behavioral aspects of certain species or the need to explore the spatial distribution or density of a species to understand habitat needs. Economic geography can broadly be understood as the study of the relationship between the production of goods and services (labor, finances, manufacturing, and raw materials) and consumption of those goods and services. Many types of human-animal relations are rooted in processes of economic exchange— whether through market systems where animal products are bought and sold for money or through subsistence economies where people are raising or hunting their own animals for survival. Ethical geographies are concerned with examining how notions of right and wrong not only differ spatially, but are also place dependent (i.e., it is considered ethically right to experiment on animals in a laboratory but not in a home). Regarding humans and animals, numerable ethical issues arise. For example, is keeping animals in captivity or conducting medical experiments upon them right or wrong? Is eating other species right or wrong? Finally, political geography studies the ways in which individuals and collectives (as groups or states) negotiate (or not) competing views of how societies should be organized and how social organization differs across scales, space, and places. The politics of human-animal relations can be very contentious, resulting in acts of violence or the introduction of laws that ban or allow certain practices like circuses or dog racing, or they can become mired in difficult decisions about which animals or humans should be where.

The last point to consider is how geographers do their research—in other words, what methodologies do they use. As with the concepts and categories, research methodologies depend on the types of questions you are asking and the types of data you need to collect to answer your questions. In many cases, geographers will focus on gathering either qualitative (detailed, in-depth case studies) or quantitative (generalizable/statistical) data. Sometimes research questions will involve the need for a combination of depth and breadth. One research method, archival research, involves gathering and reviewing primary sources. These sources could be diaries, government documents, newspapers, photographs, and so on. Archived materials are usually of a historical nature, but increasingly geographers are using archives of electronic media (blogs, websites, videos) to gather first-person data. A second method is field research. This term encompasses a spectrum of techniques that all involve the geographer going out "into the field" or the real world to collect data. Data can be gathered through interviews, surveys, visual studies, the collecting of environmental data (e.g., species counts, land transects), or ethnographic (long-term local studies) data. A third method is cartography and related mapping sciences. In addition to traditional mapmaking, GIS allows geographers

THE TIGER: *PANTHERA TIGRIS*

Tigers are in the order Carnivora and the family Felidae. Thirty-seven species of cats exist in the world today, seven of those considered big cats: tigers, lions, leopards, jaguars, snow leopards, clouded leopards, and cheetahs. Of the eight subspecies of tigers, three are extinct, and the other five are considered highly endangered. Experts estimate only a few thousand wild tigers remain and worry that wild populations are becoming too low to be genetically viable. The cat family emerged around forty million years ago, and tigers have evolved over the past two million years to succeed in a wide variety of habitats—from the bitterly cold and snowy areas of eastern Russia to the dense tropical rain forests of Southeast Asia.

The Amur or Siberian tiger is the largest of the subspecies, and a large male can be ten feet long and weigh between four hundred and six hundred pounds. Tigers are strong swimmers and highly evolved predators. The coat of the tiger with orange or white background under black stripes is not only unique to the species, but each tiger has its own unique pattern. Tigers stalk their prey, waiting to get close enough to pounce and crush the neck with their jaws. They are solitary animals unless it is mating season or a mother is raising cubs. Tigers can breed throughout the year, and females gestate two to three cubs for about one hundred days. The cubs stay with the mother for nearly two years before going off on their own. Social contacts are maintained through an intricate scent system that serves to demarcate territory as well as determine breeding females. Males will compete to control access to females.

Humans have never domesticated tigers, though they have been tamed and bred for a variety of uses—circuses, zoos, medicinal parts. Humans have long had a fascination with tigers, seeing them as emblems of power, majesty, grace, and prowess. Humans have hunted tigers, and sometimes tigers have chosen to hunt humans. Today more tigers are estimated to be in captivity as pets in the United States than in the wild, a testament to our contradictory relations with them. We love them and want to save them for their wildness, yet we also want some of that wildness for ourselves, and so keeping them as pets or using them for medicine becomes a way to connect with their power. How many tiger emblems, mascots, and characters can you think of?

to combine large quantities of environmental and human data into layers on a digital map while the use of satellites for remote sensing can give physical geographers a more nuanced view of environmental properties like changes in forest cover.

Let us now use two different examples of human-animal relations to get a better glimpse of how all of these aspects of geography (branches, concepts, analytic categories, and methodologies) can come together. First, let's take a practice that involves animals—industrial agriculture. While chapter 5 will focus on this topic in depth, we can define industrial animal agriculture as a practice of modern farming that uses advanced technologies in breeding and raising livestock to produce more animals faster, in confined animal feeding operations (CAFOs). What types of research questions might a geographer ask? Where are the CAFOs (space)? How many animals are in CAFOs (scale)? What are the environmental impacts of CAFOs (human-environment)? Where and when did the transition to CAFOs occur (place)? Why (politics, economics, culture)? What of the livestock themselves (physical)? Are CAFOs an ethical way to treat our food animals (ethics)? How do the meat and dairy products get to our plates (economics)? How does the industrial animal landscape look different from more traditional livestock farming (landscape)? What do farmers, animal rights activists, the general public, or doctors feel with regards to CAFOs, or how would we study the experience of the animals themselves (methodologies)? As you can see there are a multitude of geographic starting points. If we take the example of human interaction with a specific species—the tiger—we can also map out several different directions. Who are tigers and where are they found (physical/spatial)? What is the cultural lore around tigers (human/culture)? How are tigers commodified (turned into objects for buying and selling) for their parts or for tourism (economic)? How are conflicts between people and tigers addressed (human-environment, politics, culture)? Should tigers be kept as pets (ethics, culture, politics)? What is a tiger farm (space)? Should we try to save tigers (ethics, politics, place)? You will have an opportunity to think more deeply about these topics over the next several chapters, but now you must begin to see what it means to think geographically and how it can help us understand our world more thoroughly.

OVERVIEW OF ANIMAL STUDIES

As we have already seen Jennifer Wolch and Jody Emel (1998) point out—animal geography did not emerge in a vacuum. Nor are geographers the only ones with an academic interest in human-animal relations. Just as animal geography is a subfield to geography, it is also a subfield of the multidisciplinary field of human-animal studies (HAS), which has been rapidly developing over the past twenty-five years. The origins of academic work are usually credited to philosophers Peter Singer and his book *Animal Liberation* (1975) along with Tom Regan's book *The Case for Animal Rights* (1983). Both of these books brought to light the treatment of different animals within a modern, Western context. This treatment was often shocking for people to confront. For many, reading these books was the first time they learned about industrial animal agriculture or what went on behind many laboratory doors. Singer mapped out a utilitarian view of animals that argued that society should have as its goal maximizing good and minimizing harm—something he argued humans were not doing when it came to animals. Regan took a different direction for his explanations and argued that animals, like humans, are "subjects of a life" and, as such, have intrinsic value and the right to live their lives as subjects. As HAS developed, it moved beyond these discussions, in a disciplinary sense, from philosophy to the natural and social sciences and to more subjective fields such as literature and the arts. Figure 1.3 shows HAS course offerings in the United States. According to the Animals and Society Institute, HAS courses are also offered in Australia, Canada, Germany, Great Britain, Israel, Poland, and New Zealand in many different academic disciplines. Nine major HAS journals focus on humanities, social sciences, critical

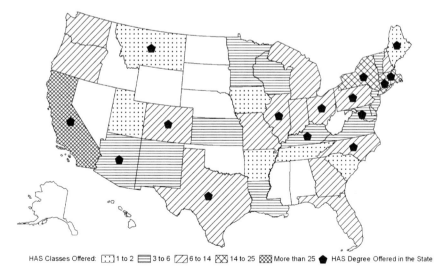

HAS Classes Offered: ▒ 1 to 2 ▤ 3 to 6 ▨ 6 to 14 ▩ 14 to 25 ▦ More than 25 ⬟ HAS Degree Offered in the State

Figure 1.3.   Map of HAS Course Degree Offerings in the United States. *Source:* **Data for this map was compiled from the Animals and Society Institute Course database and is current as of 2011.**

animal studies (CAS), law, and hard animal sciences (see the resources section at the end of the chapter).

Ken Shapiro, the founding editor of the journal *Society and Animals*, argues that the emergence of animal studies in the academy echoes the rise of other identity-based studies such as ethnic, gender, and race studies programs (Shapiro 1993). This parallel development occurs as social movements work to challenge existing problematic social structures—whether legal, cultural, political, or economic—in the public realm while scholars recognize the need to recover the histories and treatment of marginalized and oppressed groups of humans. This comparison is not to claim that nonhumans or their treatment is commensurate with different human groups, but to recognize (1) certain groups have been treated differently over the course of history around the world and (2) understanding this treatment is part of understanding the collective experience of human societies. Animal studies scholars argue that animals have basically been taken for granted by humans and kept to the edges of not only our daily lives, but our intellectual thought, and we need to "recover" animals as animals and the multiple relations they have with humans. This reclaiming has proven to be no easy feat—with either animal-focused social movements or within the academy. Indeed, "animals may be the last group to be brought into the circle of morality and subjectivity; no other group has been admitted without bloodshed and strife" (Wolch and Emel 1998, xii).

What has made animal studies so difficult? Rhoda Wilke and David Inglis point to several explanations in the introduction to their edited collection *Animals and Society: Critical Concepts in the Social Sciences* (2007). They argue that the first challenge is a history of intellectual thought across academic disciplines, which has seen animals as biotic elements that can be used or not used by humans. The cementing of animals as objects for human use in the academy has been traced to René Descartes, a sixteenth-century philosopher/scientist. He concluded animals could not speak and lacked souls and therefore were no more than instinctually driven automatons. While Descartes applied his famous dictum—"I think, therefore I am"— to humans, he did not apply it to animals. We have, in the vast majority of intellectual thought since his time, continued to render animals as mute background noise to human action. In essence, we have had the "perception that animal matters are relatively trivial in comparison to what has been historically defined as the central objects of social-sciences—namely humans" (Wilke and Inglis 2007, 5).

Descartes, however, is not the only impediment to taking animals and human-animal relations seriously in the Western academic tradition. Descartes was building on a long Christian tradition of separating humans and animals. The Bible states that God gave man dominion over all the other species and that only man is made in the image of God. The ramifications of this construction of human superiority are seen today in the concepts of speciesism and anthropocentrism—both contributing another major hurdle for animal studies. *Speciesism*, a word coined by British psychologist Richard Ryder in the mid-1970s, is an attitude that regards all nonhumans as inferior to humans and therefore not part of a human moral system. We can understand speciesism as akin to concepts such as racism or sexism where one group of people is made out to be less than other groups simply because of who they are. An anthropocentric (human-centered) view does not necessarily see all nonhumans as inferior, but believes that what we choose to do or not do to other species matters only in terms of how it will impact humans. These two concepts promote a view of humans that emphasizes their uniqueness in relation to all other forms of life, and that uniqueness translates into intrinsic superiority. "Just as much contemporary scholarship recognizes that forms of chauvinism in particular thought-systems can result in unfounded derogations of particular human groups, the critique of anthropocentrism asserts that speciesism is an intellectually unjustified way of thinking about life on earth" (Wilke and Inglis 2007, 7).

This leads into a third point contributing to the difficulties of animal studies—that of interrogating the problems inherent in studying human perceptions and constructions of animals versus the animals themselves. Even talking about humans *and* animals, which has been done during the course of this chapter, is problematic to a certain degree. Humans are, in fact, animals and animal studies scholars walk a line between the utility of framing the topic as human-animal relations and reifying the separation of humans from other species. HAS scholars and animal geographers continue, however, to use the terms *humans* and *animals* or *human-animal relations* because of the fundamental ease of phrasing. We will continue to do that in this book as well even as we tease out the human(s) and the animal(s). We will also continue to recognize the difficulty of trying to speak about other species while knowing they cannot reciprocate in the same way. A last point complicating the field of animal studies is the issue of scholarship versus advocacy. Academic scholarship is supposed to be objective and without a political agenda. Many critics of animal studies argue that the field is just a front for animal rights activism and therefore not really scholarship. Animal studies scholars themselves have to be wary of the ways in which their own views may impact their research projects. Some animal studies scholars see no problem with academic advocacy: indeed, a whole subfield of critical animal studies (CAS) professes a radical animal rights agenda. CAS returns us to the links between HAS and other group studies scholarship. Challenging the status quo is often threatening to those who benefit from it, and the development of gender studies, critical race studies, and ethnic studies have had/do have very similar tensions.

OVERVIEW OF THIS BOOK

From this overview of geography and HAS we can begin to understand where animal geography is coming from and how it fits into geography and the larger animal studies community. As the purpose of this book is to *introduce* animal geography, it is designed to be a synthesis and survey of animal geography rather than an in-depth theoretical analysis. Furthermore, the use of nonanimal geography HAS material has been curtailed in order to focus exclusively on animal geography insights. The remainder of the book is organized to move from the history of animals in geography to the "new" animal geography today and how it is being used to understand human-animal relations in different places. Chapter 2 focuses specifically on the development of geographic work on animals by conducting a broad survey of historical accounts of animals and then outlining the development of modern-day animal geography. Chapter 3 explores the animal geographies of the home through the examples of pets and culture. Chapter 4 moves to the places of working animals and includes discussions of zoos, laboratories, and other working animals (e.g., police dogs, elephants). Chapter 5 examines the geographies of farmed animals by comparing and contrasting pastoral (herding), industrial, and modern alternative practices of raising animals for their meat, eggs, and milk as well as for their fur. Chapter 6 brings human relationships with wild animals into focus and explores the places of encounter such as in urban areas, at ecotourism spots, while out hunting, and during human-wildlife conflict. Chapter 7 provides a summary of the book and concluding thoughts about the future directions of animal geography both within and outside the discipline of geography. Furthermore, each chapter ends with keywords, review discussion questions, an assignment related to the material, and a list of resources and references to help you focus on the main points and have some guidelines for where to go to move deeper into individual topics.

DISCUSSION QUESTIONS

1. Why do geographers and animal studies practitioners argue that it is important to study human-animal relationships?
2. How does the broad field of animal studies differ from the biological sciences' focus on animals?
3. What are the key points about geography that will be used to frame the discussion of animal geography? Brainstorm examples besides tigers and CAFOs to see what types of geographically based questions you can come up with.

KEYWORDS/CONCEPTS

animal geography
anthropocentrism
critical animal studies (CAS)
human-animal studies (HAS)
human geography
human-environment geography
landscape
mapping sciences

region
physical geography
place
power
scale
space
speciesism

PRACTICING ANIMAL GEOGRAPHY

Brainstorm as many different types of human-animal relationships that you can think of (e.g., zoos, petting zoos, roadside zoos, rodeos, eating flesh, wearing animals) and then create your own human-animal "map" that shows your "animal autobiography." Reflect on the results.

RESOURCES

Animal Geography Specialty Group of the Association of American Geographers: http://www.animalgeography.org
Animal Law Review: http://law.lclark.edu/law_reviews/animal_law_review
Animal Planet: http://animal.discovery.com
Animals: http://www.animal-journal.eu
Animals and Society Institute: http://www.animalsandsociety.org/main
Antennae: http://www.antennae.org.uk
Anthrozoös: http://www.bergpublishers.com/?TabId=519
Association of American Geographers: http://www.aag.org
Between the Species: http://digitalcommons.calpoly.edu/bts
Humanimalia: http://www.depauw.edu/humanimalia
Institute for Critical Animal Studies: http://www.criticalanimalstudies.org
Journal of Animal Ethics: http://www.press.uillinois.edu/journals/jane.html
Journal of Animal Law: http://www.animallaw.info/policy/pojouranimlawinfo.htm
National Geographic: http://www.nationalgeographic.com
Society and Animals: http://www.animalsandsociety.org/static/resources-publications ?tcid=45&_i=sub

REFERENCES

Beston, Henry. 1992. *The Outermost House*. New York: Henry Holt.

Clottes, Jean, and Valérie Féruglio. 2011. "The Cave of Chauvet-Pont-D'Arc." Accessed June 6. http://www.culture.gouv.fr/culture/arcnat/chauvet/en/index.html.

FitzGerald, Michael. 2010. "Inside the Great Inanke Cave." *Wall Street Journal.* Accessed June 6, 2011. http://online.wsj.com/article/SB10001424052748704509704575019501899388186.html.

Gregory, Derek, Ron Johnston, Geraldine Pratt, Michael Watts, and Sarah Whatmore, eds. 2009. *Dictionary of Human Geography*. Malden, MA: Wiley-Blackwell.

MSU. 2011. "Michigan State University Animal Legal and Historical Center 'World Materials.'" Accessed March 3. http://www.animallaw.info/nonus/index.htm.

Muller, Tom. 2011. "Secrets of the Colosseum." *The Smithsonian*, January, 1–4.

Regan, Tom. 1983. *The Case for Animal Rights*. Berkeley: University of California Press.

Shapiro, Kenneth J. 1993. "Editor's Introduction to *Society and Animals*." *Society and Animals* 1 (1): 1–4.

Singer, Peter. 1975. *Animal Liberation: A New Ethics for Our Treatment of Animals*. New York: Random House.

Steinfeld, Henning, Pierre Gerber, Tom Wassenaar, Vincent Castel, Mauricio Rosales, and Cees de Haan. 2006. *Livestock's Long Shadow: Environmental Issues and Options*. Rome: United Nations Food and Agriculture Organization.

Thomas, David S. G., and Andrew S. Goudie, eds. 2000. *Dictionary of Physical Geography*. 3rd ed. Malden, MA: Wiley-Blackwell.

Wilke, Rhoda, and David Inglis, eds. 2007. *Animals and Society: Critical Concepts in the Social Sciences*. New York: Routledge.

Wolch, Jennifer, and Jody Emel, eds. 1998. *Animal Geographies: Place, Politics, and Identity in the Nature-Culture Borderlands*. New York: Verso.

Chapter Two

A History of Animal Geography

Each year *National Geographic* compiles a list of the top ten weirdest species that were discovered that year. In 2010, the list included the *Tyrannobdella rex*—a leech with large teeth from Peru, a purple octopus from Canada, the so-called Yoda bat from Papua New Guinea, a new species of snub-nosed monkey in Myanmar, a pink handfish in Australia, and the "Simpsons" toad in Colombia ("Ten Weirdest New Animals" 2010). These terrestrial and aquatic species span the globe and stand out for their unique attributes. The leech uses its teeth to saw into the orifices of mammals (all of them!) while the "Yoda bat" not only looks as wise as its *Star Wars* namesake but also has tubular nostrils that point outward and away from each other. The pink handfish uses its fins to walk across the ocean floor rather than swim and the "Simpsons" toad has a long, pointy snout that resembles the villain Mr. Burns from *The Simpsons* television show. In 2009, the list included such notables as the blob fish, the vegetarian spider, pygmy seahorses, and the giant wooly rat. That such lists are being compiled today can surprise people who think that all the species in the world have already been discovered by science; however, we are quite a ways from accomplishing that mission. In fact, new animal species are discovered by scientists on a regular basis and brought into societal awareness. Estimates are that anywhere from five million to fifty million different species (all life-forms) can be found on earth, yet only around 1.8 million of these have been named and classified by taxonomists (Hickman et al. 2011).

Scientific classification was first developed by Swedish biologist Carl Linnaeus in the eighteenth century and has been subsequently modified into a system that divides organisms into eight major taxa: domain, kingdom, phylum, class, order, family, genus, and species. This system allows scientists to map not only the various organisms that exist in the world, but also the relationships among organisms. Animals, from the Latin for "having breath," are a kingdom broadly defined as life-forms, either vertebrate or invertebrate, that consume and digest their food rather than photosynthesize or absorb it (Hickman et al. 2011). Of all known species, about 75 percent (or approximately 1.3 million) are classified as animals. Of the animals, about 73 percent are insects and only around 3.9 percent are vertebrates. Of that 3.9 percent, only 9 percent are mammals. Amphibians, reptiles, birds, and fishes constitute the rest of the vertebrate category (Groombridge and Jenkins 2002). The distribution of animal species is uneven across the planet's major biomes—or geographic areas with similar climate, flora, and fauna. Generally, greater numbers of animal species live toward the equator because of plant productivity and climate stability, and the top five countries with the most classified mammal species are Indonesia, Brazil, China, Mexico, and Peru (International Union for the Conservation of Nature [IUCN] 2011).

While knowing something about the blob fish or the Yoda bat might make you a favorite in social situations, that knowledge also puts you in line with the long lineage of humans who have tried to classify the other beings that we share the planet with. Indeed, the Western scientific methods of taxonomy are only one way of ordering animal species. Indigenous peoples the world over have their own methods (Howitt and Suchet-Pearson 2003). Traditional indigenous knowledge (TIK), from the Aborigines in Australia to the Inuit of North America, organizes animals based on use (e.g., food bird, nonfood bird), according to religion (e.g., totemic or sacred animal), or by attributes (e.g., land, water, and air) (Rose 1996; Waddy 1988). These systems tend to reflect much more relational categorizations of human-animal interaction and are limited to the species known to the area in which the indigenous group lives. As with the discussion of different definitions of "animal" in chapter 1, cultural differences in classification remind us that our ordering of the nonhuman animal world is not singular but rooted in geographically based social identities.

With geography's history of cataloging the wheres and whys of life on earth, that animals—their types, their distributions, their habitats—have been a topic of study should come as no surprise. Yet *how* geographers have studied other animals has evolved over time, and documenting this evolution is the goal of this chapter. We will move chronologically through history and arrive at the "new" animal geography of today. While a chronological approach can appear to selectively skim over the many wider cultural and political contexts, in this case we will (1) focus specifically on animals within geographic writing and (2) briefly introduce you to this history. We will examine the development of three "waves" of animal geography. Others have delineated the notion of "waves" in animal geography but only in a cursory way (Wolch, Emel, and Wilbert 2003), so here we are going to fill in more of the history with a survey of geographic writing on animals prior to the turn of the twentieth century. As we move through time you will want to pay attention to *how* animals are being written about, *which* animals get covered, and *which* human-animal relationships are being considered. As we will see, the current wave of animal geography expands the field of study quite dramatically from predecessors who focused on straightforward scientific cataloging of species to asking questions about human-animal interactions and the lives of individual and collective animals themselves.

ANIMALS AS OBJECTS OF HISTORIC GEOGRAPHIC FASCINATION

By highlighting a few examples from a wide sweep of geographic history, we will see that animals have not been absent. Over the past millennia geographers have reveled in noting strange species and documenting the behaviors of different cultures toward nonhumans. While the vast majority of these works take animals as natural objects (not subjects), unquestionably under human control, they are rich sources for considering the length of time certain relations have been in place and also for understanding that even then geography was essential to human-animal relationships. Ancient intellectuals were intensely curious about their world and, without the benefits of modern scientific tools, they did their best to explain what they saw and experienced around them.

Herodotus (484–425 BCE), a Greek historian, wrote *The Histories* about the origins of the Greco-Persian wars and combined that history with systematic descriptions of other lands and peoples. Commentary on human-animal relations abounds in his works. For example, he writes:

> In whatever house a cat dies of a natural death, all the family shave their eyebrows only; but if a dog dies, they shave the whole body and the head. All cats that die are carried to certain sacred houses, where, being first embalmed, they are buried in the city of Bubastis. (Herodotus 1885, 121)

Herodotus later comments in the same work:

> Thus, then, as far as the lake Tritonis from Egypt, the Libyans are nomads, eat flesh, and drink milk, but they do not taste the flesh of cows, for the same reason as the Egyptians, nor do they breed swine. Indeed, not only do the women of the Cyrenaeans think it right to abstain from the flesh of cows, out of respect to Isis in Egypt, but they also observe the fasts and festivals in honor of her: and

the women of the Barcaeans do not taste the flesh of swine in addition to that of cows. These things, then, are so. (300)

Bubastis, a city in Lower Egypt, was the center of worship for the goddess Bast or Bastet. She was seen as a protectress and usually depicted with a cat or lion's head and a woman's body. Cats were very important to the ancient Egyptians for their hunting abilities, keeping grain stores safe from rodents, and mother cats were also seen as very affectionate and protective of their kittens. Tombs in Bubastis were filled with mummies of deceased cats. Isis, one of the principal gods of ancient Egypt, represented the ideal mother as well as being seen as the matron of nature and magic. She was often depicted with a cow's head since cows were seen as nurturers and life-givers because of their milk. Herodotus also describes many different types of animals and where they could be found and how local people hunted them or otherwise interacted with them. Herodotus summarily laid the groundwork for the centuries of descriptions of animals and human-animal relations to come.

Eratosthenes (285–205 BCE), the person who coined the word *geography*, is a second key Greek figure. While no complete copy of his three-volume work *Geographika* exists, we do have fragments in which we see the attempt to combine an understanding of the nature of the surface of the earth with human-inhabited portions. Thus we get descriptions of places like India, about which he comments:

Almost the same animals appear in Aithiopia [area of Ethiopia today] and throughout Egypt as in India, and there are the same ones in the Indian rivers except the hippopotamus, although Onesikritos says that this horse is also there. The people in the south are the same as the Aithiopians in color, but in regard to eyes and hair they are like the others (because of the moisture in the air their hair is not curly). Those in the north are like the Egyptians. (Roller 2010, 84)

As contact between different ancient cultures and landscapes increased over time, more comparisons could be made in terms of similarities and differences between places and peoples. The descriptive cataloging would continue to be the norm. Strabo (63 BCE–24 CE) was also an ancient Greek geographer, most famous for his seventeen-volume *Geographica*, a compilation of information about his known world. He writes simultaneously of peoples, landscapes, and animals. Of the Iberian peoples he states: "And intermingled with their forces of infantry was a force of cavalry, for their horses were trained to climb mountains, and, whenever there was need for it, to kneel down promptly at the word of command. Iberia produces many deer and wild horses. In places, also, its marshes teem with life; and there are birds, swans and the like; and also bustards in great numbers" (Strabo 1988, 107). And of the Gallic peoples he writes: "Food they have in very great quantities, along with milk and flesh of all sorts, but particularly the flesh of hogs, both fresh and salted. Their hogs run wild, and they are of exceptional height, boldness, and swiftness; at any rate, it is dangerous for one unfamiliar with their ways to approach them, and likewise, also, for a wolf" (243).

What we see here would be repeated over the centuries as explorers and naturalists documented the variety of cultures and environments around the world. We can see here that the use of horses and pigs was naturalized, and no attempt is made throughout the work to ponder the origins of agriculture, or the hows and whys of pig and horse domestication or people's relations to them—their subjugation is simply a given. In this sense, Strabo's work follows that of Herodotus and Eratosthenes. While this style of documentation is necessary and helpful, in terms of animal geography today, it is really the first level of approaching human-animal relations.

Alexander von Humboldt (1769–1859), a German naturalist and explorer, traveled extensively in Latin America at the turn of the nineteenth century. An excellent writer, Humboldt captured people's imaginations even as he detailed in catalog form the physical geographies he encountered. A couple of quotes from his book *Aspects of Nature* (1850) will serve as examples not only of his style, but also of the worldviews of European colonial explorers at the time. On the ox and the horse he comments:

> The ox and the horse have followed man over the whole surface of the globe, from India to Northern Siberia, from the Ganges to the River Plate. . . . The ox wearied from the plough reposes, sheltered from the noontide sun in one country by the quivering shadow of the northern birch, and in another by the date palm. The same species which, in the east of Europe, has to encounter the attacks of bears and wolves, is exposed in other regions to the assaults of tigers and crocodiles. (39)

Here we see the recognition of humans' impact on animals via domestication, but also the notion that humans—and animals—experience domestication differently in different parts of the world because of the local biogeographies. Note that he constructs domestication as a process whereby the oxen and horses seem to simply "follow" humans—almost as if they were choosing to work for humans. In another chapter on the nocturnal life of animals in the primeval forest, he documents his experience along the Rio Apure in what is present-day Venezuela:

> Soon after 11 o'clock such a disturbance began to be heard in the adjoining forest, that for the remainder of the night all sleep was impossible. . . . There was the monotonous howling of the aluates (the howling monkeys); the plaintive, soft, almost flute-like tones of the small sapajous; the snorting grumblings of the striped nocturnal monkey (the Nyctipithicus trivirgatus, which I was the

first to describe); the interrupted cries of the great tiger, the cuguar or maneless American lion, the peccary, the sloth, and a host of parrots, or parraquas, and other pheasant-like birds. . . . If one asks the Indians why this incessant noise and disturbance arises on particular nights, they answer, with a smile, that "the animals are rejoicing in the bright moonlight, and keeping the feast of the full moon." To me it appeared that the scene had probably originated in some accidental combat, and that hence the disturbance had spread to other animals, and the noise had increased more and more. (212–213)

With eloquent prose, Humboldt epitomizes the manner of approaching animals as either exotic natural objects instinctually responding to external stimuli or as objects of man's use during this period of European exploration and colonization.

Another early forerunner to animal geography was American George Perkins Marsh (1801–1882). Often considered the first environmentalist, Marsh is best known for making the case that humans can have a tremendous (and detrimental) impact on the environment if they are not careful. However, in a book written after his extensive travels, he documents, in what is perhaps the first geographic work to focus on an individual nonhuman species, the camel and how it might be a useful species to bring to the United States (Marsh 1856). For Marsh, "the ship of the desert has navigated the pathless sand-oceans of the Sahara, and of Gobi, and thus not only extended the humanizing influences of commerce and civilization alike over the naked and barbarous African and the fur-clad Siberian savage, but, by discovering the hidden wells of the waste and the islands of verdure that surround them, has made permanently habitable vast regions not otherwise even penetrable by man" (22). In the book he covers the two species of camels, their physiology, their temperament, their distribution, and their adaptations, along with providing a variety of descriptions of the ways in which different human groups used the camel historically and in his time. While he tries to make the case for introducing the camel into the United States as a utilitarian beast of burden and a military animal, that vision was not to be implemented over the long term; however, his work comes closest to what animal geography would develop into nearly one hundred years later because of his ability to combine the varied human-animal relations with a more focused effort on the camel as a subject rather than an object.

THE FIRST WAVE OF ANIMAL GEOGRAPHY

As geography became formally institutionalized in the academy in the late nineteenth century, the study of animals was considered a key part of the dis-

THE CAMEL: *CAMELUS DROMEDARIUS* AND *CAMELUS BACTRIANUS*

The one-humped dromedary of southwestern Asia and the two-humped Bactrian camel of eastern Asia are part of the taxonomic family Camelidae of which there are six species: the llama, the alpaca, the guanaco, the vicuña, the dromedary, and the Bactrian camel. The camel family originated in North America around forty million years ago and only recently (approximately two million years ago) spread to South America and Asia. They are the principal herbivorous mammals of arid habitats. Today, the vast majority of camels are dromedaries, and the Bactrian camel is endangered in the wild in Asia.

Camels are normally six to seven feet tall at the hump and weigh between one thousand and fourteen hundred pounds. They have a two-toed foot that rests on a rugged sole pad and a split upper lip that they can move somewhat independently. They can consume a wide variety of plants and go for long periods without water. The hump is used for fat storage, which also enables them to go for a long time without food. To conserve water they produce very little urine and very dry feces and raise their body temperatures to reduce water loss through sweating. They can close their nostrils and their long eyelashes to keep out blowing sand. The gestation period is around thirteen months and the babies are walking and able to keep up with their mothers shortly after birth. Their social organization is normally in herds with a dominant male and his harem. They are both intelligent and often ornery and make a variety of sounds that communicate their state of being.

The dromedary was probably domesticated between 2,000 and 10,000 BCE in southwest Asia while the domestication of the Bactrian camel probably occurred around the same time, but separately in central Asia. Camels were domesticated for use as transport and food, making the survival of nomadic cultures in arid environments possible. Camels are also used in entertainment: camel wrestling (between two male camels) and camel beauty contests are common in countries like Saudi Arabia and Turkey. In Western popular culture Joe Camel of Camel cigarettes (named for the iconic species from Turkey where R.J. Reynolds got its tobacco from) became a magnet for controversy in the late 1980s through the 1990s because of his "cool" character possibly influencing children to smoke.

cipline and came to be known as zoogeography—from the Greek for "animal-land description/writing." Now seen as the first wave of the field of animal geography, zoogeography was defined as "the scientific study of animal life with reference to the distribution of animals on the earth and the mutual influence of environment and animals upon each other" (Allee and Schmidt 1937). The animal life referred to did not include humans. The field was heavily influenced by the work of Charles Darwin (1859) and Alfred Russel Wallace (1876), through their theories of natural selection and evolution, and Phillip Sclater, through his pioneering work, which divided the world into six zoogeographic regions (Sclater 1858; see also Darlington 1966). Indeed, as Wilma George notes in her book *Animal Geography*, "before Darwin and Wallace announced the theory of natural selection it was generally assumed that each species lived in the region best suited to it because it had been especially created for that place. Sloths were created for South America, elephants for Africa and India, and rats presumably for the whole world" (1962, 33). Hence the reason that more causal or relational questions about animals were never really asked. That the world didn't always work out like this (e.g., the spread of the rabbit in Australia or the European starling in the United States, demonstrating that no clear reason could be given as to why they should not have lived there before) meant that a plethora of questions could be asked about animals once they had been cataloged. The vast majority of this work regarded animals as natural objects to be studied separately from humans with a goal to "establish general laws of how animals arranged themselves across the earth" (Wolch, Emel, and Wilbert 2003, 185). Zoogeography was even seen by Richard Hartshorne, a prominent American geographer who spent the majority of his career at the University of Wisconsin, as a key part of his vision of geography as the study of areal differentiation (Hartshorne 1939). For Hartshorne, studying the conflux of human and physical systems in regions made geography unique, and zoogeography was an essential systematic subfield.

Zoogeographers cataloged species and their current and historical distributions and also studied how the environment influenced species' adaptations. One of the earliest, and perhaps the only, case of actual animal experimentation in geography was a study by V. E. Shelford (1903) on the relationship between evaporation and its effects on warm-blooded animals. He conducted a series of experiments on frogs, salamanders, millipedes, spiders, and insects by placing them in glass tubes and running air of different evaporating powers through the tubes until the animals died. He found that forest animals died more quickly in drier air while sand dune animals died more quickly in moist air, thereby demonstrating how animals were adapted to, and impacted by, their natural habitats.

The early "20th century was marked by the focus of attention on details, on studies that traced the evolution and movement of animals through both space and time," and one atlas and two books on animal geography published during this time exemplify the zoogeographic approach (Stuart 1954, 445). In 1911, J. G. Bartholomew, W. Eagle Clarke, and Percy H. Grimshaw published their *Atlas of Zoogeography*. A collection of over two hundred color maps based on the zoological regions of Sclater and Wallace illustrates the spatial distributions of known species at the time along with briefly cataloging and describing individual species. One example of such description is the following: "In the Oriental region the family [Simiidae] is represented by the Orang-Utan (*Simia satyrus*). This animal is characterized, like the Gorilla, by its very large canine teeth, and its brain approaches that of man more closely than does that of any other ape. At the present day the Orang is restricted to the dense primeval forests of Borneo and Sumatra, but in the Pliocene period it appears to have inhabited northern India" (Bartholomew, Clarke, and Grimshaw 1911, 13). They emphasized that the geographic approach was key because, unlike faunal catalogs or ecological studies, zoogeography tried to understand why the same physical area such as a tropical rain forest could result in very different arrays of species between Brazil and West Africa.

In 1913, Marion Newbigin, a prominent Scottish geographer, and a lecturer in zoology and biology at the Edinburgh School of Medicine for Women, and one of the founders of modern British geography, published *Animal Geography*. She recognized that geographers had made great strides in studying the world's plants, and she felt that animals should receive the same consideration. In her view, animal geography should recognize that the animals of natural regions form part of the features of regions, and, therefore, attention to animals should be directed to the adaptations of the animals to their environment rather than relationships between species. In her book she moved methodically through the main biomes of the world, from the tundra to caves and the sea, describing the animal life that could be found. The descriptions, like other animal geography works we've seen, still read more like a catalog:

> There is nothing very characteristic about the reptiles of the savannas and hot deserts, but of the numerous lizards which occur there one or two may be named as showing interesting peculiarities. In the sandy districts of Western and Southern Australia occurs *Moloch horridus*, a lizard covered with spines and tubercles. It lives upon ants, and has the curious power of being able to absorb water through its rough skin. This is presumably an adaptation to permit the animal to avail itself of the rare showers which fall in the desert region where it lives. (Newbigin 1913, 142)

W. C. Allee and Karl P. Schmidt's book, *Ecological Animal Geography* (1937), is their rewritten translation of German geographer Richard Hesse's (1924) book *Tiergeographie auf oekologischer Grundlage* (Animal geography on an ecological basis). Concerned that so little about animal geography and animal-environment relations was being published in English, they set out to follow the same structural format as Newbigin's in terms of cataloging the different animals through the biomes, but they also argued that "we have had an over-supply of travel which yielded animal pelts and alcoholic material; we need rather observations on the relations between animals and their environment" (Allee and Schmidt 1937, xiii). To remedy this they added the additional species and species-environment relations that had been discovered since Newbigin's book and included detailed chapters on topics such as the conditions of existence for animals, barriers to distribution, the effect of geographic isolation, and the influence of extent of range. They aimed to move beyond more straightforward cataloging and to really work toward scientific theorizing of how animals lived and adapted to different conditions (see figure 2.1).

Furthermore, Allee and Schmidt (1937) included a brand-new chapter— "The Effect of Man on the Distribution of Animals"—foreshadowing the second wave of animal geography. In this chapter Allee and Schmidt describe the ways in which "civilized man" impacts other species in the cases of deforestation, managed forests, agriculture, gardens and parks, buildings, unintended transport, and direct eradications. They conclude by commenting:

> So great have been the changes in the vegetation and animal life of the world with the spread of civilized man, that over wide areas the natural phenomena of geographic zoölogy and of ecology in general are completely secondary, approachable from the agricultural or economic standpoint rather than from the biological. . . . The only hope for the preservation of natural conditions for the future, in temperate latitudes, and probably in the tropics as well, lies in the establishment of state and national parks, in which primitive conditions are maintained, to serve as refuges and sanctuaries for wildlife. (556)

Other geographers during this time commented on animals but as a side story to human differences. For example, several interesting discussions take place in *The Principles of Human Geography* by Paul Vidal de la Blache, who is considered the father of French geography. In a chapter on transportation he discusses the role of the draft animal in human history and how the "various types developed in widely different environments were useful as a match for the variety of obstacles to be surmounted" (quoted in Martonne 1926, 354). He claims the ox may have been the first beast of burden since it appears in Chaldean, Chinese, and Germanic mythologies, but that open

The size difference of the ears of the desert fox (*Canis zerda*), of the European fox (*C. vulpes*) and of the polar fox (*C. lagopus*) is shown in Fig. 112. In Siberia, the ears of the wild hogs, the red deer, the roe deer, the fox, and wildcat are relatively smaller, often positively smaller, than those of the smaller German forms.[48] Figure 113 exhibits the extreme difference in the surface de-

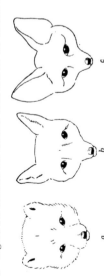

Fig. 112.—Head of arctic fox (*Canis lagopus*), *a*; red fox (*Canis vulpes*), *b*; and desert fox (*Canis zerda*), *c*.

velopment and body form which may be found in antelopes from cold and warm climates. Similar differences appear in *Gazella picticauda*, of the Himalayas at an elevation of 4000-5000 m. above sea level, and *G. bennetti*, inhabiting the plains of north and central India; with approximately equal length, the mountain animal has shorter legs, ears, and tail.[49]

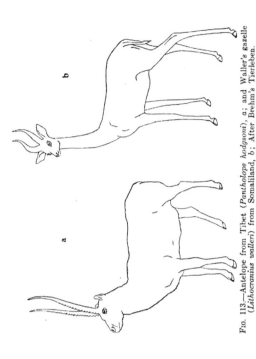

Fig. 113.—Antelope from Tibet (*Pantholops hodgsoni*), *a*; and Waller's gazelle (*Lithocranius walleri*) from Somaliland, *b*; After Brehm's Tierleben.

Figure 2.1.   Images from *Ecological Animal Geography* by Allee and Schmidt (1937). *Source:* Author's personal collection.

countries like plains or deserts needed different species like the camel and horse. That the Bactrian (two-humped) camel was docile, had a strong sense of direction, and could find its own food made it a suitable species for domestication; however, he also noted that "it is not a fighter, its sluggish habit could not be altered without damage, [and] expeditions have resulted in a veritable slaughter of these unfortunate animals" (quoted in Martonne 1926, 357). Apparently domestication has its limits! He also claims that donkeys must have been domesticated in Upper Egypt at the dawn of history because the monuments show that they were plentiful some time ago.

A second way animals appear in such a text is during Paul Vidal de la Blache's discussions of the various peoples around the world. In his description of the "central European type" he makes a point of recognizing the role of hogs in this culture: "Hogs still wallow in village streets; they live with the peasants; their fattening is an object of affectionate concern; their sacrifice an important date in the rural calendar. . . . And it was not otherwise in the days when Gallic hams figured in Roman gastronomy, or when ancient texts referred to the innumerable droves of swine roaming about in 'glandiferous Pannonia' [a province of the Roman Empire]" (quoted in Martonne 1926, 223). This type of discussion directly echoes the ancient Greek geographers. So what we have during the first wave of modern animal geography is a focal emphasis on zoogeography but also an obvious secondary interest in human-animal interactions.

THE SECOND WAVE OF ANIMAL GEOGRAPHY

By the middle of the twentieth century, fields such as biology and zoology began taking on more and more of the traditional cataloging of animal species, their distributions, and their ecologies. This development left geographers with an interest in animals to begin to focus more on human-animal relations in place and space. This second wave of animal geography, while never entirely relinquishing zoogeography, saw a rising interest in the impact of humans on wild animals and in human relations with livestock. With respect to wild animals we can turn to Michael Graham and Fraser Darling for two examples. Graham (1956) documents the rise and fall of the whaling industry around the world and argues that the near extinction of some whales was a harbinger of what could happen to other wild species if humans could not effectively regulate their use of sea creatures as a natural resource. Darling also takes an instrumental view of animals and argues that "man advances materially and ultimately in his civilization by breaking into the stored wealth of the world's natural ecological climaxes" (1956, 778). In his overview of the history of pastoralism he comments on the extensive ways in which humans control the natural world and reshape landscape. He cites as an example of wild pastoralism the plains Indians of the United States who used fire to burn prairie and extend the range of the wild bison that they relied upon so heavily.

Geographers interested in animals also turned to the "Berkeley School" in California led by Carl Sauer. For Sauer, understanding *how* humans transformed the environment from "natural landscapes" to "cultural landscapes" was what constituted geography. This study of the cultural landscape as cultural ecology—how human cultures shape and are shaped by their environment—necessarily involved addressing the issue of human-animal relations primarily through studying domestication. In influential books like *Seeds, Spades, Hearths and Herds* (1952), Sauer examined how animal domestication helped create cultural landscapes because the use of animals necessitated grazing areas, holding pens, livestock feed, and so on, and did cause humans to significantly alter local environments. This new cultural ecology approach was helpful in introducing the idea that human culture has a huge role to play in terms of human-animal relations. For example, Sauer was never convinced that domestication was a purely survival-mode phenomenon; he raised the role of economics, religion, and perhaps even feelings of empathy or kinship toward other species as contributors to domestication processes. While this definitely demarcated a huge change from the position of animals as natural objects in zoogeography, this view overall retained a sense of human separation and dominance over other animals. Furthermore, Sauer never attempted to get at the experiential lives of the animals themselves and remained content with descriptions of their physical features and their use by humans.

In 1960, geographer Charles F. Bennett (1960, 1961) had just begun his long career at the University of California, Los Angeles, and he was concerned about the loss of traditional zoogeography to other academic disciplines and called for a "cultural animal geography" and a renewed interest in teaching about animals as part of the geography curriculum. He wanted to "impart to Geographers and their students, a needed appreciation of the importance of animals as an element in the landscape as well as an awareness that man is himself an animal and is intimately involved with the whole panoply of biological phenomena" (1960, 14). He recognized the three existing types of geographic studies on animals at the time: faunistic (the study of spatial distributions), historical (the study of geologic and fossil evidence), and ecological (the study of environmental dynamics shaping animal distributions). Building on the work of Sauer and other cultural ecologists examining animal domestication, Bennett called for a fourth type and named it "cultural animal geography." He envisioned it as a field that would "accumulate, analyze, and systematize data relevant to the interactions of animals and human cultures" (1960, 13). His examples of topics for research included humans as an agent of species dispersal and subsistence hunting and fishing. One must recognize that for the cultural ecologists, animals were only one part of what they were studying; Bennett was calling for a specialty area within geography that brought together zoogeographic elements with the recognition that humans and animals impact each other in multiple ways.

The next three decades of geographic research on animals did delve into cultural animal geography, yet it was not specifically called cultural animal geography but instead cultural ecology or cultural history. A few examples will demonstrate this cultural ecology approach that emphasized domesticated animals and landscape changes. Daniel Gade's (1967) work on the guinea pig serves as an excellent example. Gade explains that the guinea pig is one of a very few mammals to have been domesticated in the New World—along with the llama, alpaca, dog, and Muscovy duck. The guinea pig is more closely related to the chinchilla and porcupine than to a rat or mouse, and South America provides a home for seventeen different species of guinea pigs. Gade documents the variety of ways people in this part of the world have used guinea pigs and suggests that domestication may have first taken place because they were scavenging off of human settlements and over the years became domesticated and used as food, in medicine, and also as pets. Mummified guinea pigs have even been found buried with humans. Peasants used this animal as a meat source but also for larger, ceremonial purposes. They usually lived (and still do) with humans and have free run of houses. They also continue to be used in folk medicine as the fat is considered good for tight nerves and earaches, while warm viscera are applied to the body for rheumatic and abdominal pains.

For an excellent example of a second-wave focus on a cultural animal geography of a particular species we can turn to Frederick and Elizabeth Simoons's *A Ceremonial Ox of India* (1968). "The approach in this study, it should be emphasized, is not that of the zoologist, but that of the cultural geographer. A balanced view of the mithan will be attempted by following two methods adapted from more general geographic studies. One involves consideration of the mithan by topic. The other treats the mithan within each of several ethnic groups where it is found" (1968, xiv). The mithan is a domesticated bovine related to guar and banteng cattle. Distinguished by its docile demeanor, it lives in the highland regions of northeast India, Myanmar, and Bhutan. Over the course of this text they outline the physiology of the mithan, its distribution and habitat, its role in several different cultures, its role in local economies, and the general history of domestication. They find that while the mithan is used across this region, it is used in different ways. Some communities raise it only for ceremonial purposes, while others use the animals as stores of value, or for dowry payments or conflict resolution.

Continuing to build on the complex cultural connections between humans and animals, while still within an anthropocentric framework, Grossman (1984) documents how cultural variables like values and local practices can be crucial forces shaping present-day animal-related livestock practices. In other words, human use of nonhumans as livestock is not defined only by economic utility. He does this through a case study of sheep in Papua New Guinea. Sheep are not native to this area, and in trying to argue for the economic development of Papua New Guinea, the government and the international community promoted sheep production; however, in spite of abundant land and labor, sheepherding did not "take" in the region because sheep were not seen as "high" enough or valuable enough entities for ceremonial exchange—local people preferred their own animals. Thus, he concludes that culture plays a role in human-animal agricultural relations.

Frederick Simoons and James Baldwin (1982) take a different direction with agriculture and human-animal relations by focusing on gender—specifically women and the breastfeeding of animals. While most people might find this practice bizarre, in essence it is no different than humans' drinking the breast milk of cows or goats. In their review of the global literature on the practice of women breastfeeding other species they were following through on a speculation by Carl Sauer that this practice might be a link in the history of domestication. They found documented histories of breastfeeding animals everywhere except for China, Europe, the Near East, and Africa. They concluded that there are four overlapping reasons for the practice: affection (one example was of aborigines bringing home wild babies, especially dingo pups that were raised and cared for with emotion), economics (such as raising dogs for hunting or pigs to be eaten), ceremonial (to provide an animal for sacrifice and ritual as did the Ainu of Japan who would raise wild-caught baby bears in cages for years in order to sacrifice them), and human welfare (meaning the health interests of a mother or child). They speculate that a fifth reason might be more symbolic—to show that an animal is being brought in as a member of the group, in essence building a kinship bond like family.

G. S. Cansdale's book *Animals and Man* (1952), while before Bennett's call, is the other main example of the second wave of descriptive cultural animal geography during the middle of the century. Cansdale, at the time the superintendent of the London Zoological Gardens, focused specifically on the ways in which humans and animals interacted in a variety of situations. With chapters on competition and conflict, the introduction of foreign animals by humans, the role of animals in service to man, and other areas such as sports, captivity, and religion, his book was the first one to so thoroughly demonstrate the intersections in human-animal lives. While the style is mainly that of cataloging the myriad relations, almost as if he were cataloging species, the inclusion of cultural intersections beyond domestication reveals the larger potential of cultural animal geography. For example, figure 2.2 shows a visual study of how animals are used in everyday culture (banks and coats of arms)—highlighting the symbolic place of animals in human culture—a concept that will come to greater prominence with the third wave of animal geography.

THE THIRD WAVE OR "NEW" ANIMAL GEOGRAPHY

The third wave or "new" animal geography has, thus far, built more upon Sauer, Bennett, and Cansdale than on the zoogeographic tradition. As we learned in chapter 1, the impetus for this "new" animal geography in the mid-1990s was the intersection of events in the world, an academic reassessment of culture and subjectivity, and a desire to unpack the "black box" of nature (Wolch and Emel 1998). In 1995, a special topics issue of the journal *Environment and Planning D: Society and Space* edited by Jennifer Wolch and Jody Emel laid the groundwork for this re-visioning of what constituted animal geography. The special issue was followed by their edited book *Animal Geographies* (1998), which included revised papers from the journal, and then a second anthology of research articles titled *Animal Spaces, Beastly Places*, edited by Chris Philo and Chris Wilbert (2000). The 2000s saw an increase in animal geography publications in major geography journals, and in 2009 Monica Ogra and Julie Urbanik founded the Animal Geography Specialty Group of the Association of American Geographers (see resources section at the end of this chapter for a list of major geography journals).

Philo and Wilbert define this new animal geography as a subfield that "focus[es] squarely on the complex entanglings of human-animal relations with space, place, location, environment and landscape" (2000, 4). In addition, they state that animal geography works to explore

> the conjoint conceptual and material placements of animals, as decided upon by humans in a variety of situations, and also to probe the disruptions of these placements as achieved on occasion by the animals themselves. It is to record many of the animal spaces specified in one way or another by humans, but also to spy on something of the beastly places made by animals themselves, whether wholly independent of humans or when transgressing, even resisting, human spatial orderings. (24)

The arms of Barclays Bank   The Great White Horse, at Ipswich

(*Below*) The arms of Martins Bank and the arms of the Royal Veterinary Colleg∠

Picture Post Library

Figure 2.2.   Images from *Animals and Man* by Cansdale (1952). *Source:* Author's personal collection and reprinted with permission of Random House and Getty Images.

This mission statement obviously sounds nothing like animal geography as zoogeography, but it does sound similar to a cultural animal geography/cultural ecology approach; however, we have already discussed how the cultural animal geography that came to be from the 1950s through the early 1990s focused almost exclusively on domesticated animals under human control. Two features distinguish the "new" animal geography (hereafter "animal geography") from the first two waves: (1) an expanded notion of human-animal relations beyond the domesticated livestock to include *all* locations of human-animal encounters (e.g., zoos, labs, pets, popular culture) and (2) attempts to bring in the animals themselves as subjects of their own lives—whether part of ours or not—instead of just as objects of human control. A more straightforward definition of animal geography then might be "the study of where, when, why and how nonhuman animals intersect with human societies." Animal geographers rely on the concepts (space, place, scale, landscape, power), categories of analysis (cultural, economic, ecological, ethical, political), and methodologies (archival, field research, mapping sciences) of human geographers that we explored in chapter 1. Because these concepts and categories of analysis so easily blur together when looking at specific topics, the remaining chapters are organized around the major arenas in which humans and animals encounter each other—home, work, farm, and wild—to demonstrate the complexity of geographic thinking with even the most deceptively simple examples such as pets.

Before turning to these arenas we first need to get a handle on several fundamental linkages between geographic concepts and categories with respect to human-animal relations that are foundational to animal geography. The first linkage is that of the relationships between place (the unique), power (control or contesting of), cultural identity (gender, ethnicity, race, religion, etc.), and animals. In a foundational article for animal geography, Glen Elder, Jennifer Wolch, and Jody Emel (1998) outline the ways in which practices on animal bodies (e.g., eating, fighting, researching) are used to police relations between different human groups in different places. They point out that what is an accepted animal practice depends on the dominant human group in a particular place, and animal geographers can study how actions (1) reinforce human-to-human power relations and (2) reinforce boundaries between humans and other species. They cite as one example the predominantly Hispanic practice of horse-tripping, a sport where riders on horseback lasso the legs of a running horse thereby "tripping" it. In some cases this can cause harm to the horses, but in Hispanic cultures the ability to do this is seen as a great skill. In the United States, this practice has been constructed as largely unacceptable by the dominant Anglo culture, which has an affinity for horses. Those who practice horse-tripping argue that the Anglo culture is being hypocritical because horses are often horrifically hurt in any number of "acceptable" horse sports in the United States such as racing, steeple-chasing, barrel racing, or cross-country events. What are we to make of this? Elder, Wolch, and Emel want us to understand that animals "appear to be one site of struggle [power] over the protection of national identity [place] and the production of cultural difference [culture]" (1998, 72). Furthermore, "norms of legitimate animal practice are neither consistent nor universal" (73). We can also consider the relationship between culture, place, and power with human-animal relations in a more symbolic manner. Lesley Instone (1998) points out that even animal symbolism is place based: for example, the coyote, constructed as a trickster figure in Native American traditions and also in modern social theory, does not translate as such around the world to places where people do not know this animal. Coming from an Australian context she argues that the symbolism of the dingo —the Australian wild dog—makes more sense within her worldview because she knows that species of dog. Regardless of what the animal symbol is being used for, Instone argues that it needs to have a place affiliation to be properly understood and worked with. Therefore, we must understand that not only are humans working out relations with animals, but human groups are also competing, confronting, and conforming with each other about animals *in addition to* having relations to them.

A second set of foundational relations to consider is that of the analytic category of ethics, and the conceptual categories of place and power. Bill Lynn's (1998a, 1998b) concept of geoethics provides a way to think through issues of right and wrong and how we determine how we should live and interact with other species. Lynn challenges more traditional value paradigms such as anthropocentrism (humans are what matters), biocentrism (living beings matter), and ecocentrism (both living and nonliving systems along with their interactions matter) by arguing that they are too rigidly ideal to be useful in sorting out human-animal relations. He calls for instead recognition of three points: (1) Geography constitutes all ethical problems. In other words where you are matters for whether an action is conceived of as right or wrong. (2) As geographic beings we all are a part of the earth. (3) We are all part of a geographic community. What sets the concept of geoethics apart is that it recognizes both the whole (ecosystem) and the parts (individuals), in essence constructing a value paradigm with plural centers of moral value. For Lynn this difference means a moral principle of equal consideration can be developed without that necessarily meaning equal treatment; so while we might not choose to treat pigs the same as humans we should still consider them. In fact we need to develop place-sensitive evaluations of moral problems. For example, is it morally right to eat animals or not eat animals? An anthropocentric approach might say yes, because human happiness is all that matters, while a biocentric approach might be concerned about the animals being eaten and the ecocentric approach concerned about the impact on ecological systems. A geoethical approach would ask us to consider both the individual animal and the larger ecosystem. Where and how is the animal being raised and slaughtered? What is the impact of that animal's life on the ecosystem compared to the scale of animals produced for consumption? These types of questions allow a more nuanced approach than trying to fit all human-animal relations into one value system.

This brings us to the notion of responsible anthropomorphism as a way to link culture, ecology, and place (Johnston 2008). For Catherine Johnston the "new" animal geography "demanded the re-politicization of animals as bodies and voices, not merely ideologies or conceptual tools" (634). Building on anthropologist Tim Ingold's concept of dwelling, which emphasizes the ways in which humans are always in the midst of experiencing the world, she argues that we must approach human-animal relationships in the same way, and thus, her notion of responsible anthropomorphism is "a way of knowing about and knowing with animals not based on our shared sentience, our shared place in the world, or any other such abstract philosophical argument, but on our actual relationships, our day-to-day living and working" (646). While her notion can be considered part of nonrepresentational theory, "an umbrella term for diverse work that seeks to better cope with our self-evidently more-than-human, more-than-textual, multi-sensual world" (Lorimer 2005, 83), because of her emphasis on interaction, Johnston echoes Lynn by contextualizing the place-based ways we do (or do not) interact with other animals. In so doing she is trying to push to the front the recognition that the experience of being alive is one that happens among all species, and once we recognize that, we can, perhaps, move into different relationships based on our shared dwelling.

Johnston's responsible anthropomorphism also dovetails with work by Owain Jones (2000) on the "ethics of encounter." Like Lynn, Jones rejects ethical paradigms that focus exclusively on the normative and generalizable and emphasizes species or other groupings of animals rather than individual animals. He states that "the ethical invisibility of the individual non-human other has been and remains extremely useful and probably essential to modern societies" (279). An ethics of encounter requires humans to actually look at, or encounter, the individual animals humans have (in)direct relations with, then to recognize our shared dwelling, and finally to conduct a more balanced ethical inquiry that sees the individual animal, the individual human, and the larger whole. For all three of these animal geographers, the ethics of different human-animal relationships must be examined in the particular places in which they happen; this examination is the only way to make the individual animals visible *and* to open oneself to the global complexity of human-animal relations.

This discussion takes us to the last major foundational concept of the "new" animal geography, that of hybridity. As we have moved into a period of posthumanism where the notion of the human as the only subject that matters, or as really the only subject, has been shown to be false by fields such as biology with work on genetic engineering pushing the boundaries of species, or by work in ethology pushing the boundaries of what we understand about animal intelligence and animal cultures, we now recognize that human identities are created not in isolation but in relation to other living beings and inanimate things (Castree and Nash 2006). This idea has come into animal (and human) geography from the field of science and technology studies and actor-network theory (Murdoch 1997). Actor-network theory (ANT) has attempted to dissolve the strict boundaries that came out of the Enlightenment, the Scientific Revolution, and modernism (see chapter 1), which constructed the world as a series of purified dualisms such as society/nature, subject/object, and human/animal. We can understand a dualism as a pair of relations where one relation is valued above the other and treated accordingly. Historically, only the valued half of the pair has been seen as "subjects" or "actors" shaping the world. Actor-network theorists argue that, akin in some ways to dwelling, humans react and respond to the animate and inanimate world around them, much more intimately than had been realized previously. They argue that, in fact, the world cannot be separated into these dualisms because the actors that make up the world include much more than humans. In this way of looking at the world, nonhuman entities are not passive objects, but also actively constitute and create the world.

The concept of hybridity emerged from ANT and has been used in animal geography to understand the ways humans, animals, and human-animal relationships are cocreated (Whatmore 2002). Hybridity, like ANT, recognizes that agency—the ability to act or effect change—is multidirectional and does not come from humans alone. Hybridity adds to ANT by emphasizing that individuals are never really purified entities—that everything is engaged in relations. In essence, we are all hybrids, entities that are some type of mix between heretofore separate categories. We can be hybrids between humans and technology by having an artificial heart or talking on a cell phone. We can be hybrids between humans and animals by our very biology and our ability to absorb a pig heart instead of an artificial one. From an animal geography perspective, the specific geographies of hybrids are of most interest. For example, Sarah Whatmore (2002) points out that a human-animal relationship with a captive zoo elephant differs from a human-animal relationship with a free-roaming wild elephant, and differs still again from one with a circus elephant. In all cases, *which* humans and *which* elephants matter because the humans and elephants morph and shift with each encounter (whether you are a trainer, an owner, a spectator, a baby elephant born into a zoo, or an elephant captured in the wild and forced into captivity). Hybridity, then, helps animal geographers to see that (1) subjects exist—human, elephant—in and of themselves, and (2) a constellation of identity relations forms when different human-human, human-animal, and animal-animal configurations appear in specific places (Lulka 2009).

While hybridity really allows animal geographers to focus on the individual relationships, I would like to expand on the phrase "power geometry" coined by geographer Doreen Massey (1993) to "interspecies power geometry" to capture the idea that our relations with nonhumans are hybrid, yes, but the *power* within the relations is very different—something the concept of hybridity does not fully capture. Massey focuses on understanding the ways in which different social groups and individuals are positioned within unfolding processes of globalization and argues that "it is not simply a question of unequal distribution, that some people move more than others, some have more control than others. It is that the mobility and control of some groups can actively weaken other people" (62). We must, therefore, not only understand the flows and movements of globalization, but also understand how globalization is working to the power advantage of certain groups of people over others differently in different places. Extending this notion to our understanding of human-animal relations is key because it allows us to see not only how differential hybrid relations develop and exist, but also how processes of power work differently depending on the place of the relation. So my power relationship with a zoo elephant gives me more control because the elephant is enclosed, but in the wild the elephant would have the power over me because of its physical strength.

We have seen over the course of this chapter on the chronological history of animals in geography (1) an ongoing interest in nonhumans and (2) a clear development of a subfield of animal geography that has gone through its own metamorphosis. The first wave of zoogeography focused on the distributions and adaptations of mainly wild animals. Humans were largely out of the picture. With the second wave, built on the cultural ecology vision of Carl Sauer, we saw a focal point on human relations with domesticated animals. As we can see from this last chapter section, the animal geography that we end up with today deepens the first two waves of zoogeography and cultural ecology. Does this mean zoogeography and animal cultural ecology have disappeared from the discipline? No, indeed many geographers continue to study animals from these perspectives and provide crucial data on distributions and ecosystem relations and nonindustrial livestock practices.

The "new" animal geography, however, distinguishes itself by decentering the human as the focal subject, recognizing the agency of nonhumans, and demanding a geographically rich analysis of the ways in which the full spectrum of human-animal relations come into being, exist, evolve, and disappear. Concepts such as geoethics, hybridity, interspecies power geometry, and responsible anthropomorphism provide the foundation for our study of animal geography because they get to the heart of untangling the dynamic intersections of place, ethics, culture, and identity across the human-animal relations spectrum. Animal geography argues that regarding humans as the pivot of the world no longer makes sense and that we are so deeply intertwined with other species, the only way we can understand ourselves is to understand them and our relations to them. The next chapter will take us into our first focal area—that of the home and culture. Human-animal interactions via pets, especially in the West and industrialized countries, are often the most direct and intimate human-animal relationships people have, but as we will see, an animal geography perspective brings a revealing level of complexity.

DISCUSSION QUESTIONS

1. When and where do you encounter individual animals versus groups of animals or their parts? How does your behavior change depending on the location? Why?
2. What are the reasons for the development of the three "waves" of animal geography in the modern period? What differentiates them?
3. Where and how have you learned about animals in your life? Which "wave(s)" does this learning correspond to?
4. Brainstorm examples of human-animal relations and consider how practicing responsible anthropomorphism via geoethics or an ethics of encounter might change these relations.

KEYWORDS/CONCEPTS

cultural ecology
dualism
ethics of encounter
geoethics
hybridity

interspecies power geometry
"new" animal geography
object/subject
responsible anthropomorphism
zoogeography

PRACTICING ANIMAL GEOGRAPHY

1. Go back to your animal autobiography and reframe it as an "interspecies power geometry." How does this change your own context? Have you added or removed any animals? Why?
2. Make a list of the weirdest species in your area and see how many of your friends and family can identify these animals. What did you find? Why do you think this is the case? How does this exercise lead us to consider which animals (if any) we might have a responsibility to consider?

RESOURCES

Annals of the Association of American Geographers: http://www.aag.org/cs/publica tions/annals

Antipode: http://www.wiley.com/bw/journal.asp?ref=0066-4812

Applied Geography: http://www.journals.elsevier.com/applied-geography

Area: http://www.blackwellpublishing.com/journal.asp?ref=0004-0894

Economic Geography: http://www.clarku.edu/econgeography

Environment and Planning A & D: http://www.envplan.com

Geoforum: http://www.journals.elsevier.com/geoforum

Global Environmental Change: http://www.journals.elsevier.com/global-environ mental-change

Human-Wildlife Interactions Journal: http://www.berrymaninstitute.org/journal/ index.html

International Journal of Geographical Information Science: http://www.tandf.co.uk/ journals/tgis

Journal of Applied Ecology: http://www.journalofappliedecology.org/view/0/index.html

Journal of Biogeography: http://www.blackwellpublishing.com/journal .asp?ref=0305-0270

Landscape and Urban Planning: http://www.journals.elsevier.com/landscape-and -urban-planning

Political Geography: http://www.journals.elsevier.com/political-geography

Professional Geographer: http://www.aag.org/cs/publications/the_professional _geographer

Progress in Human Geography: http://phg.sagepub.com

Transactions of the Institute of British Geographers: http://onlinelibrary.wiley.com/ journal/10.1111/%28ISSN%291475-5661

There are also numerous regionally focused journals.

REFERENCES

Allee, W. C., and Karl P. Schmidt. 1937. *Ecological Animal Geography: An Autho-rized, Rewritten Edition Based on* Tiergeographie auf oekologischer Grundlage *by Richard Hesse*. New York: John Wiley & Sons.

Bartholomew, J. G., W. Eagle Clarke, and Percy H. Grimshaw. 1911. *Atlas of Zooge-ography.* Edinburgh: John Bartholomew and Co.

Bennett, C. F. 1960. "Cultural Animal Geography: An Inviting Field of Research." *Professional Geographer* 12 (5): 12–14.

———. 1961. "Animal Geography in Geography Textbooks: A Critical Analysis." *Professional Geographer* 13:13–16.

Cansdale, G. S. 1952. *Animals and Man.* London: Hutchinson.

Castree, Noel, and Catherine Nash. 2006. "Posthuman Geographies." *Social and Cultural Geography* 7 (4): 501–504.

Darling, F. Fraser. 1956. "Man's Ecological Dominance through Domesticated Animals on Wild Lands." In *Man's Role in Changing the Face of the Earth,* edited by William L. Thomas, Jr., 778–787. Chicago: University of Chicago Press.

Darlington, Philip J. 1966. *Zoogeography: The Geographical Distribution of Animals.* New York: John Wiley & Sons, Inc.

Darwin, Charles. 1859. *On the Origin of Species by Means of Natural Selection, or the Preservation of Favoured Races in the Struggle for Life.* London: John Murray.

Elder, Glen, Jennifer Wolch, and Jody Emel. 1998. "Le Pratique Sauvage: Race, Place, and the Human-Animal Divide." In *Animal Geographies*, edited by Jennifer Wolch and Jody Emel, 72–90. New York: Verso.

Gade, Daniel W. 1967. "The Guinea Pig in Andean Rolk Culture." *Geographical Review* 57 (2): 213–224.

George, Wilma. 1962. *Animal Geography.* London: Heinemann.

Graham, Michael. 1956. "Harvests of the Seas." In *Man's Role in Changing the Face of the Earth,* edited by William L. Thomas, Jr., 487–502. Chicago: University of Chicago Press.

Groombridge, B., and M. D. Jenkins. 2002. *Global Biodiveristy: Earth's Living Resources in the 21st Century.* UNEP-World Conservation Monitoring Center. Cambridge, UK: Hoechst Foundation.

Grossman, Lawrence S. 1984. "Sheep, Coffee Prices, and Ceremonial Exchange in Papua New Guinea." *Geographical Review* 74 (3): 315–330.

Hartshorne, Richard. 1939. *The Nature of Geography: A Critical Suvery of Current Thought in the Light of the Past.* Lancaster, PA: Association of American Geographers.

Herodotus. 1885. *Herodotus: A New and Literal Version, translated by Henry Cary.* New York: Harper and Brothers.

Hesse, Richard. 1924. *Tiergeographie auf ökologischer Grundlage.* Jena, Germany: Gustav Fischer.

Hickman, Cleveland P., Larry S. Roberts, Susan L. Keen, David J. Eisenhour, Allan Larson, and Helen l'Anson. 2011. *Integrated Principles of Zoology.* 15th ed. New York: McGraw-Hill.

Howitt, Richard, and Sandra Suchet-Pearson. 2003. "Ontological Pluralism in Contested Cultural Landscapes." In *Handbook of Cultural Geography*, edited by Kay Anderson, Mona Domosh, Steve Pile, and Nigel Thrift, 557–569. Thousand Oaks, CA: Sage.

Humboldt, Alexander von. 1850. *Aspects of Nature, in Different Lands and Different Climates; With Scientific Elucidations*, translated by Mrs. Sabine. Philadelphia: Lea and Blanchard.

Instone, Lesley. 1998. "The Coyote's at the Door: Revisioning Human-Environment Relations in the Australian Context." *Ecumene* 5 (4): 452–467.

International Union for the Conservation of Nature. 2011. "Geographic Patters of Diversity." Accessed June 1. http://www.iucnredlist.org/initiatives/mammals/analysis/geographic-patterns.

Johnston, Catherine. 2008. "Beyond the Clearing: Towards a Dwelt Animal Geography." *Progress in Human Geography* 32 (5): 633–649.

Jones, Owain. 2000. "(Un)ethical Geographies of Human–Non-human Relations: Encounters, Collectives and Spaces." In *Animal Spaces, Beastly Places: New Geographies of Human-Animal Relations*, edited by Chris Philo and Chris Wilbert, 268–291. New York: Routledge.

Lorimer, Hayden. 2005. "Cultural Geography: The Busyness of Being 'More-than-Representational.'" *Progress in Human Geography* 29 (1): 83–94.

Lulka, David. 2009. "The Residual Humanism of Hybridity: Retaining a Sense of the Earth." *Transactions of the Institute of British Geographers* 34:378–393.

Lynn, William S. 1998a. "Animals, Ethics, and Geography." In *Animal Geographies*, edited by Jennifer Wolch and Jody Emel, 280–297. New York: Verso.

———. 1998b. "Contested Moralities: Animals and Moral Value in the Dear/Symanski Debate." *Ethics, Place and Environment* 1 (2): 223–242.

Marsh, G. P. 1856. *The Camel: His Organization Habits and Uses Considered with Reference to His Introduction to the United States*. Boston: Gould and Lincoln.

Martonne, E., ed. 1926. *Principles of Human Geography by Paul Vidal de la Blache*. London: Constable.

Massey, Doreen. 1993. "Power-Geometry and a Progressive Sense of Place." In *Mapping the Futures: Local Cultures, Global Change*, edited by Jon Bird, Barry Curtis, Tim Putnam, George Robertson, and Lisa Tickner, 59–69. London: Routledge.

Murdoch, Jonathan. 1997. "Inhuman/Nonhuman/Human: Actor-Network Theory and the Prospects for a Nondualistic and Symmetrical Perspective on Nature and Society." *Environment and Planning D: Society and Space* 15:731–756.

Newbigin, Marion. 1913. *Animal Geography: The Faunas of the Natural Regions of the Globe*. Oxford: Clarendon.

Philo, Chris, and Chris Wilbert, eds. 2000. *Animal Spaces, Beastly Places: New Geographies of Human-Animal Relations*. New York: Routledge.

Roller, Duane W. 2010. *Eratosthenes' Geography: Fragments Collected and Translated, with Commentary and Additional Material*. Princeton, NJ: Princeton University Press.

Rose, D. B. 1996. *Nourishing Terrains: Australian Aboriginal Views of Landscape and Wilderness*. Canberra: Australian Heritage Commission.

Sauer, Carl. 1952. *Seeds, Spades, Hearths and Herds*. New York: American Geographical Society.

Sclater, P. L. 1858. "On the General Geographical Distribution of the Members of the Class Aves." *Journal of the Proceedings of the Linnean Society* 2:130–145.

Shelford, V. E. 1903. "The Significance of Evaporation in Animal Geography." *Annals of the Association of American Geographers* 3:29–42.

Simoons, Frederick J., and James A. Baldwin. 1982. "Breast-Feeding of Animals by Women: Its Socio-cultural Context and Geographic Occurrence." *Anthropos* 77:421–448.

Simoons, Frederick J., and Elizabeth S. Simoons. 1968. *A Ceremonial Ox of India: The Mithan in Nature, Culture and History.* Madison: University of Wisconsin Press.

Strabo. 1988. *The Geography of Strabo, v. 2, with an English Translation by Horace Leonard Jones.* London: William Heinemann.

Stuart, L. C. 1954. "Animal Geography." In *American Geography: Inventory and Prospect*, edited by Preston E. James and Clarence F. Jones, 443–451. New York: Syracuse University Press.

"Ten Weirdest New Animals of 2010: Editor's Picks." 2010. *National Geographic.* Accessed May 15, 2011. http://news.nationalgeographic.com/news/2010/12/photogalleries/101207-top-ten-weird-new-animals-2010.

Waddy, Julie. 1988. *Monograph: Classifications of Plants and Animals from a Groote Eylandt Aboriginal Point of View.* Darwin: Australia National University.

Wallace, Alfred Russel. 1876. *The Geographical Distribution of Animals, with a Study of the Relations of Living and Extinct Faunas as Elucidating the Past Changes of the Earth's Surface.* New York: Harper and Brothers.

Whatmore, Sarah. 2002. *Hybrid Geographies: Natures Cultures Spaces.* London: Sage.

Wolch, Jennifer, and Jody Emel, eds. 1995. "Bringing the Animals Back In." Special issue, *Environment and Planning D: Society and Space* 13:735–760.

———, eds. 1998. *Animal Geographies: Place, Politics, and Identity in the Nature-Culture Borderlands.* New York: Verso.

Wolch, Jennifer, Jody Emel, and Chris Wilbert. 2003. "Reanimating Cultural Geography." In *Handbook of Cultural Geography*, edited by Kay Anderson, Mona Domosh, Steve Pile, and Nigel Thrift, 184–206. Thousand Oaks, CA: Sage.

Chapter Three

Geographies of More-than-Human Homes and Cultures

In 2007, American football player Michael Vick was found guilty of running an illegal dogfighting ring and sentenced to twenty-one months in federal prison. Evidence of violent abuse and neglect along with the rescue of several of the pit bulls found on his property created a firestorm of controversy. While some of the dogs were euthanized, others went to Best Friends Animal Sanctuary in Utah to either be rehabilitated as pets or to live out the rest of their lives in a safe environment. Vick is responsible for paying for the care of the dogs and he has also become a spokesperson against dogfighting. In 2009, Travis, a pet chimpanzee, living with his owner Sandra Herold in Connecticut, mauled Charla Nash. Nash nearly died in the attack and had most of her face ripped off. Subsequent surgeries have attempted to reconstruct her face, but she has no eyes. Travis was nearly fourteen years old, and the cause of the attack is uncertain. Travis had known Nash for several years; however, on that day she was wearing a new hairstyle and had a different car, and Travis had been given Xanax by Herold. Travis was fatally shot by a police officer responding to the 911 call. In his younger days Travis had appeared in television ads and on television shows as an entertainer. This incident, like the Michael Vick case, created a tremendous amount of publicity because of its shocking violence; however, the geographic issues these two incidents raise provide an excellent segue into our first animal geography umbrella topic— that of the place of animals in our homes and in our cultures.

For most of us the animals that we come into contact with on a daily basis are either those animals that live with us in our homes or those we encounter in our cultural contexts. Questions of place, identity, ethics, politics, and cultural practices swirl through these two examples. For example, what constitutes a pet can be a complicated question. In the United States dogs are seen as pets who should be "in the home" so the idea of fighting them and causing them injury on purpose in nonhome locations is anathema to the majority of our culture; however, people in some subcultures within the United States see their dogs as both pets, whom they love, and fighting animals and do not see a difference (while some dogfighters do see their dogs as purely moneymaking objects). Having a chimpanzee as a pet is altogether different for mainstream culture because chimps simply "don't belong" in a home with people. Yet again, others would disagree. Travis had lived with the Herolds for years with only a few minor incidents, and the local community often enjoyed having him around. Many conversations after both incidents centered on discussions of ethics: Is it right or wrong to fight dogs or keep chimps? Is there an ethical difference between keeping a dog that could maul someone and possessing a chimp? Dogfighting used to be a popular family sport in the United States, so what happened to send it underground, and where is underground? These incidents also raised questions about power and politics: Who has the right to tell other people what they can do with or to animals and why? Harking back to the example of horse-tripping in chapter 2, when and why is one cultural practice acceptable or not?

The goal of this chapter is to gain a geographic understanding of the myriad ways we can understand the role of animals as pets and as markers of culture and cultural difference. As we will see, this understanding moves way beyond simple cultural categorizations of this animal as pet and this animal as not-pet. In our discussion of pets we are going to consider both traditional pets such as dogs and cats and exotics such as chimpanzees. In our discussion of culture we will explore the intersections of identities, religions, ethnicities, and art to expose the links between our conceptions of animals and our practices regarding them. We will do this by moving back and forth between animal geography work and case studies outside of the current animal geography canon and by organizing the material to follow the major analytical categories of human geography—a format that we will follow for the next three chapters as well.

HISTORICAL GEOGRAPHIES

Yi-Fu Tuan's book *Dominance and Affection: The Making of Pets* (1984) is considered the immediate precursor to the third wave, or "new," animal geography because of his emphasis on exploring the role of power in the specific human-animal relationship of pet keeping. He mainly aims to demonstrate that how we treat other species in the most intimate spaces of our homes is fraught with paradox in the simultaneous practices of love and domination. For Tuan, "cruelty to animals is deeply embedded in human nature. Our relation to pets, with all its surface play of love and devotion, is incorrectly perceived unless this harsh fact is recognized" (89). When he turns to history, he sees these dual practices played out in multiple ways. On the one hand, humans have a history of seeing animals as terrible beasts, emblematic of the unknown and the uncontrollable forces of nature, but on the other hand, we also have a long history of seeing animals as emblems of beauty, power, and even the divine or spiritual. Think back, for example, to the opening of chapter 1. Did the peoples who created the zodiacs or painted animals on cave walls exemplify the view of animals as creatures to be feared or as other beings in the world? He would argue that the history of art and religion shows animals to be held in high regard, but just as with pets, humans "unhesitatingly dominate and exploit animals in myriads of ways" (72) that are linked to displays of human power over the natural world. In documenting the history of keeping wild animals—among the Aztecs, the ancient Egyptians, and even King Solomon—he says that pet keeping is simply an extension of that desire to control the nonhuman world.

Pet keeping does this by manipulating the reproductive processes of animals in order to mold them into "creatures of a shape and habit that please their owners" (95). He discusses two species at length: the goldfish and the dog. For both, he documents the length of time humans have been manipulating their genes. The Chinese have been breeding goldfish since around 950 CE, and domesticated dogs have been around in Europe for at least ten thousand years (and possibly much longer). While professing to love goldfish for their beauty, humans have also worked to breed them so that they have physical deformities such as enlarged eyes, which can easily get scraped and result in lesions (the "telescope goldfish"), or with small warts covering their bodies (the "lionhead goldfish") without regard for their individual well-being. Fish, as pets, have also become works of art and their tanks or outdoor pond areas have been designed with aesthetics and proper viewing in mind. In this way fish became part of a domesticated cultural landscape that says fish are supposed to live in tanks or small ponds.

Regarding dogs, breeding was (and is) manipulated to affect both temperament and physical features. Tuan documents how dogs have been bred with more infantile traits—droopy ears, shortened jaws, bigger eyes, smaller sizes, extended playfulness—as well as practical characteristics—the instincts to hunt, retrieve, herd, and fight—and while some of these changes have not harmed the dogs, many of them have. Dogs may have problems breathing because of the reduced nose sizes, inabilities reproducing on their own, or susceptibility to certain diseases such as hip dysplasia or eye conditions. We have so distorted the domesticated dog from its wolf ancestors that Tuan argues that this must be seen as the most brutal kind of domination. Consider what we do to pets today to make them presentable to live in our homes with us. We crop or cut off their tails and ears, remove their voice boxes, declaw them, spay and neuter them, and bathe them. All these practices suggest both the desire to control and the desire to be free from the unruly, wild side of these other species. We consider this dualist nature an example of trying to maintain the boundary separation between nature and culture even as we can see how pets are examples of hybrid entities blurring those very boundaries.

But is this dominance the only way of being with pets? He says no. The history of humans and pets is also one of deep affection and care. While the histories and evidence that we have are primarily those of the upper classes around the world, no one can doubt that many humans have deeply loved their pets even as they dominated them. Allowing pets in homes, buying them toys and furniture, and including them in family portraits demonstrate our love of these beings. Indeed, from his perspective, the expansion of modern pet-keeping practices since the late nineteenth century has come at the same time as many people have become more and more removed from the natural world, and encounters with working animals or wild animals have dramatically diminished. So while at one point in history indoor pets may have been luxuries, today they can still be luxuries, but perhaps they are also important as a way for humans to connect with the living world. Furthermore, as Rebekah Fox reminds us, "living together with another species on a daily basis necessitates a certain intimacy and recognition of individuality and personality in non-humans" (2006, 534), which provides a direct challenge to larger social systems that posit humans only dominate other animals or that humans are the only ones with the agency in pet-keeping relationships.

Two case studies by Philip Howell (1998, 2002) can help us understand historically how pet-keeping practices have become part of human societies in different ways. In the first study Howell (1998) examines the phenomenon of dog stealing in Victorian London to understand pet-keeping practices at the time as well as how these practices were part of broader social and economic conflicts between human groups. Dog stealing linked the protected space of the home with the outside world in ways that intertwined class, gender, and human-animal relations. During this time period, pet keeping was becoming a much more commodified practice with people of means being willing to pay considerable sums to purchase and care for their dogs. Dogs were not replacing children (an issue we'll see come up in the next section), but dogs as pets were becoming normalized in Victorian culture. And not just keeping dogs but also displaying affection and love for the animals were gaining acceptance. This emotional and financial investment in dogs for upper-class families made dogs an easy target for the organized crime groups that began stealing them and ransoming them for money. While dog stealing eventually

THE DOG: *CANIS LUPUS FAMILIARIS*

The domestic dog with which so many of us are familiar belongs to the taxonomic family Canidae, which covers thirty-five species in ten genera. Canids have found niches on every part of the planet except Madagascar and New Zealand and range in size from the diminutive fennec fox (2.5 to 3 pounds) to the gray wolf (up to 165 pounds). Canids originated in North America forty million to fifty million years ago during the Eocene period. Several species of wild canids are endangered: the Ethiopian wolf, with only about five hundred remaining in the wild; the African wild dog, with three thousand to five thousand; and the maned wolf, with only one thousand to two thousand.

Several traits highlight the unique attributes of canids. Their legs developed for running and they have fused wrist bones and locked front leg bones, an adaptation that prevents rotation while moving. While many canids can eat a wide variety of foods, they are predominantly carnivores and have long muzzles, strong jaws, and a combination of shearing and crushing teeth. The gestation period for most species is around sixty days with females giving birth to litters of two to six pups. Canids are pack animals and live in complex social groups that normally include a dominant male and female. Some species will hunt and raise young cooperatively. Due to life in a group, social behaviors (greeting, grooming, vocalizations) are well developed. A canid's sense of smell is much stronger than a human's. Whereas we humans have about five million scent receptors in our noses, a canid, depending on the species or domestic breed, can have anywhere from 125 million to 300 million receptors.

The earliest evidence we have for the domestic dog comes from archeological sites in Germany (from around fourteen thousand years ago) and Iran (approximately eleven thousand years ago). The domesticated dog derived from tamed wolves, and hence its classification as a subspecies of wolf. Most scientists who have studied the origin of dogs believe that dogs were probably domesticated way before these dates (new genetic technologies are pushing the date closer to a hundred thousand years ago), which makes them a logical choice for the first domesticated animal. We have no way of knowing the exact reasons for domestication, but they probably involved some combination of affection, protection, garbage consumption, or even food source. Dogs like the chihuahua (indigenous to Mexico) have been bred for hundreds of years. Today, the American Kennel Club (AKC) recognizes over 150 different breeds. New breeds—especially hybrids such as a labradoodle (Labrador-poodle) or the cheagle (chihuahua-beagle)—are becoming more popular as so-called designer breeds. While the AKC does not yet recognize these breeds, the hybrids' popularity does demonstrate people's comfort with genetic manipulation and the desire to create dogs that "fit" with what people want them to be.

was controlled by the police, Howell argues that this period in Victorian history reveals three key points about emerging modern pet-keeping practices. Firstly, dog stealing exposed the links between the home and the market, highlighting perhaps the start of pet keeping as a fully capitalist practice. Secondly, it exposed the vulnerability of the home to outside forces because while dogs were seen as property, the high level of emotional attachment to the animals revealed an "inversion of social hierarchies" (42). Where the wealthy class was normally the economic exploiter, it now found itself dominated and exploited *because of* attachment to dogs. Finally, dog stealing also revealed a glimpse into the gendered relations of pet keeping. Women were distraught at the capture of their beloved animals; however, the wealthy men were emasculated by their inability to either protect their homes or relinquish their own emotional attachments to the animals. The result actually reinforced a domestic ideology that confined women, like their pets, back into the supposed safety of home spaces.

The second case also comes from Victorian England, but explores the rise of pet cemeteries. "Idealized for its quasi-familial virtues, and inseparable from the imaginative landscape of the family, the domestic dog was firmly installed at the heart of the respectable Victorian household, a kind of household god" (Howell 2002, 8). This "household god" already made the families vulnerable to dog stealers, but with pet cemeteries, this attachment to dogs posed a challenge to religious views and the rising industrial age. "Pet cemeteries should not be seen as simply an extension of middle-class humanitarian concerns, up to and well past the line of anthropomorphic whims: rather, we should note that their proponents were attempting to redraw the boundaries of the moral community by raising the treatment of dead pets to something that approximated the treatment of dead people" (12). The clergy at the time did not believe that animals had souls and thought it was wrong to put animals with humans because they were clearly inferior. This seemingly excessive attachment to pets, dogs especially, also ran counter to the mechanistic, industrial, and scientific age where "nature" existed to be cataloged or scientifically manipulated for industrial purposes. Furthermore, the drive for pet cemeteries came mainly from bourgeois women, not only challenging their role in public, but also bringing a certain feminine emotionality into public spaces; heretofore, only mourning the loss of a person had been deemed appropriate, and mourning and commemorating the loss of a nonhuman pet challenged social norms and the depth of what constituted the moral community. As a result, according to Howell, "the dog has become, in death as in life, the first person to greet us on the other side: an entirely appropriate construction given that the dog's relationship with humans has always located it on the boundary between wildness and domesticity" (18). In both of these cases from Victorian England we can see how deeply embedded these nonhuman companions have become in the places of our homes and hearts.

What of the historical geographies of animals in larger cultural practices? Third-wave animal geographers have had little to say here thus far, but we do need to map out some of the ways animals and culture intersect as a whole to understand how our individual attitudes toward nonhumans are mediated by large social constructs. One area in which we can consider these larger social constructs is through religion. Religions are ways of making sense of the universe and humanity's role in it. If we look at the top three religions by followers—Christianity, Islam, and Hinduism—we can get a sense of how religion influences our attitudes toward, practices regarding, and experiences with animals. In the Christian tradition, both through the Bible and Christian art, animals appear often as themselves or as symbols of Christ or representations of good and bad. The belief that humans are above animals comes from the Genesis stories that state God made animals before humans and gave humans dominion—including naming—over other beings. As further separation between humans and animals, Genesis also states that only humans were made in God's image. Yet the need for humans to steward rather than dominate other species can be seen in stories such as that of Noah and his ark, in which Noah saved two of every animal from the floods that God sent to cleanse the planet. Even the story of the birth of Jesus has an interesting animal connection as Jesus was born in a barn surrounded by livestock animals—a very humble location for the son of God. Christian symbolism through animals includes the linking of Christ with a lion and the snake as the devil who tempted Adam and Eve in the Garden of Eden. Christians also celebrate saints such as St. Francis of Assisi, who is known for his love of animals, and many churches today have celebrations blessing animals in his honor. The legacy of the Christian tradition toward animals then clearly separates them from humans and celebrates human exceptionalism with a utilitarian dominion over them. Indeed, for many conservative Christians, the theory of evolution is incompatible with the Bible, and they reject the idea that humans evolved from other species, preferring to accept that God made humans wholly separate from the rest of creation.

In Islam, animals are also prominent. In the Koran, Allah created the earth and all living things, yet humans are also privileged above other species. The same notion of humans as stewards who must treat other species properly appears in Islamic tradition. In addition to the Koran, the hadiths—a collection of the stories about the Prophet Muhammad's life—give animals attention. Here we find religious decrees banning the consumption of pigs and requiring rituals such as saying Allah's name and reducing pain as much as possible when slaughtering animals. The hadiths also provide many stories of

Muhammad's kindness toward animals and an ability to see them as subjects with experiences and feelings. He often rebuked people for their poor treatment of animals—even his wife Aisha for mistreating a camel—reminding them that Allah was watching and they would be judged.

The world's third-largest religion, Hinduism, resides primarily in India and has a few very different conceptions of animals than Islam or Christianity. First of all, while Hindus do believe in one supreme deity—Brahman—they are also polytheistic, worshiping many manifestations of Brahman in lesser gods and goddesses. For example, many students worship Ganesh, the elephant-headed god of wisdom, intelligence, and education. In addition, we also find dietary restrictions in Hinduism—mainly through vegetarianism and the idea of ahimsa (nonviolence). While Hinduism does not require that followers practice vegetarianism, and indeed many Hindus do not, many others choose to as part of adhering to the principle of ahimsa that states that one should not commit acts of violence against living things. Furthermore, one of the ancient scriptures used by Hindus, the Mahabharata (a text two thousand to three thousand years old) directs Hindus to abstain from consuming flesh. Many Hindus hold the cow sacred, so much so that in major cities free-roaming cows can cause a lot of trouble! Hindu religious texts have a history of reverence for cows with Lord Krishna, a major god, often appearing as a cow, and with language that speaks of cows as "mothers" because of their ability to nurture humans with their milk. No such reverence or treatment exists for any animal in the Christian or Muslim tradition.

We see from these brief examples that religious affiliation can have a huge impact on a person's interaction with animals, shaping not only diet, but attitudes toward the relation between humans and animals, and the animals' treatment. Another facet of larger cultural practices is language. In both written and verbal forms, language is a symbolic form of communication, and animals are not simply named with language but used as metaphors and symbols in a variety of ways. For example, what does it mean to be "treated like an animal" or to be "proud as a peacock"? The English language is replete with references to animals—some based on physical characteristics, others based on superstition or happenstance, but all serving to help shape human-animal relations. Consider again what it means to be treated like an animal. We might immediately think geographically and ask which animal? Where? If you want to be treated like a celebrity's chihuahua that's probably not so bad, but if you are talking about being treated like a rabbit in a toxicity laboratory that's probably not so fun. We know, however, that with this phrase we are talking about a negative experience because no human wants to be treated "like an animal"—that is, as property, as without agency or subjectivity, as disposable, or with cruelty. But by saying humans don't want to be treated this way, are we normalizing that way of treating animals? After all, the language that we use reflects our reality, so to be treated like an animal is to be treated like you are *not* a human, a linguistic way of maintaining the strict boundaries between humans and animals.

To get more specific about species and humans we can look to animal words that we use for humans. Women can be bitches, pussies, pigs, cows, foxes, old hens, chicks, and cougars while men can be studs, dogs, and pigs—all these words signifying a particular animal. What really is wrong with a pig, or why is it so bad to be a female dog? To be "proud as a peacock" refers to the tail display of male peacocks as they show themselves worthy of a mate, yet for us this natural, practical characteristic has come to describe someone who is undeservedly showing off or being arrogant. To be a "guinea pig" means to be an experiment and normalizes the use of these otherwise pets (in a North American context) or food (in a South American context) as disposable tools of science. Animal terms within languages are used metaphorically to emphasize links between human and animal traits, but they also normalize attitudes about other species. We have serious discussions all the time about how language impacts our views of other people (think of the controversy over the word *retard* or who can use the word *nigger*) because we recognize the power of words. The same is true for our uses of language and animals. Therefore, we need to be aware of this historical baggage just like we are with the cultural baggage of religious ideas or social norms around pet keeping. In the next section, we will focus on another major category of human social organization—the economy.

ECONOMIC GEOGRAPHIES

With respect to the economic geographies of pets, very little animal geography research has been done thus far, but the existent material is promising for helping to understand and examine this category of human-animal relationship. Heidi Nast (2006a, 2006b) has pursued this avenue by arguing for "critical pet studies." For Nast, what she sees as modern-day pet love is the result of our postindustrial, postmodern society that has been influenced by global social, economic, and material processes. Indeed, "pet animals allure in part because they can be anything and anyone you want them to be" (2006b, 302). If we consider the impressive array of pet-related activities and products we can begin to absorb what she's saying. Vast numbers of television shows focus on pets, with subjects from pet care to abused and neglected animals, to celebrations of pets, like dog and cat best-of-breed contests and talent shows. A person can purchase specialized pet accessories that highlight the personality of the human—from spiked collars to pink, diamond encrusted leashes. The pet, through its human, has become quite a savvy consumer. The aisles of the big box pet stores are filled with seemingly the same number of choices for food and toys as a normal human grocery store as we seek more and more to spoil, cater to, or love (depending on your perspective) our pets.

Nast even goes so far as to argue that pets are replacing human children in today's society because "in most narcissistic contexts, child-rearing is a drag on an individual's freedom to move and consume, leading many persons to opt out" (2006a, 899). Pets, therefore, are easier to love because they are easier to fit into modern, transient lifestyles. For Nast, the result of this move away from children toward pets has a downside: "Might it be that commodified pet-animal dominance-affection-love (DAL) is a powerful means for taking human resources of time and money away from organizing activities geared toward confronting escalating inequalities and human violences locally and world-wide?" (2006b, 320). She also asks if it is "coincidental that disparities and levels of violence are increasing at the same time that we are witnessing a groundswell of pet appreciation and love?" (2006a, 901). While she does not provide any evidence that people who love their pets love other people less or are more prone to support violence against other people, she does raise some key questions about the role of capitalism in today's human-pet relations in the Western context. She wants us to consider who really benefits from the rise of modern pet love and gain a better grasp of how the pet industry manipulates pet owners and how some pet owners and non-pet owners are disadvantaged based on economics, gender, race, and so on. These issues are indeed important from a geographic perspective because they help nuance particular places: Does the location of pet stores show a class bias in that people with lower incomes have to go farther to get their pet products? Are properties being specifically taken away from lower-income areas to be turned into human-pet spaces like dog parks (see next section for an extended discussion of this topic)? Finally, what does it mean that in postindustrial societies we are seeing a rise in pet keeping alongside a decrease in child rearing? Is that really the case across all demographic groups, and why does this seem to be happening now? Are corporations really driving the pet industry, or are consumers?

The amount of money being spent on pets in the United States alone can give us an idea of the impact our love of nonhuman companions has on the economy. A geographic way to think about the relationship between pet keeping and the economy has to do with the concept of the commodity chain. Mapping a commodity chain exposes the ways in which a product is made and maintained. You don't just buy a dog from a pet store, and that's that. A commodity chain analysis would begin by asking where the dog came from.

What resources are needed to produce dogs (space, food, shelter, veterinary care), market them (advertising), sell them (pet stores), and eventually keep them (food, toys, veterinary care, training)? According to the American Pet Products Association (APPA, 2011), nearly 73 million homes in the United States contain at least one pet. What does it take to maintain the estimated 377 million animals in these homes? Pet owners spent an estimated $51 billion in 2011, which includes $18.76 billion on food, $10.94 billion on supplies, $13.01 billion on vet care, $2.13 billion on purchasing the animals, and $3.51 billion on pet services like grooming and boarding. Not an insubstantial amount of money! Furthermore, there are trends for pets just like there are trends for human clothes, hairstyles, slang, and so on. APPA notes that trends in the pet industry that are garnering more money include products that reduce a pet's carbon footprint (think recycled toys or compostable litter), pet clothes geared toward holidays, digital advances such as electric toothbrushes for dogs and digital aquarium monitors, designer foods and beauty products from such fashion houses as Paul Mitchell, Harley-Davidson, and Old Navy, and pet-friendly hotels complete with on-call dog masseuses. According to the APPA, in 2001, pet owners in the United States spent only $28.5 billion on their pets, so we can see that the past ten years have seen a dramatic increase in household spending, and the market response has also been huge as companies rush to cater to whatever people might fancy for their animals. While extrapolating these numbers globally is very difficult because of a lack of data, we can safely assume that global spending on pets is at least close to $100 billion per year.

Another avenue to understanding pet economies is to consider the global market for exotic pets. Exotic pets are those that are "out of the norm" and often captured in the wild or bred and tamed but not domesticated. Popular exotics include big cats, primates, porcupines, foxes, birds, reptiles, fish, and even kangaroos. The exotic pet industry is minimally regulated, if at all, which has led to a large global black market to feed people's desire for the strange or illegal. While people often express a deep love for their exotics, in many cases people (and also local vets) are unable (or untrained) to properly care for these animals. Quantifying the exotic pet trade today is extremely difficult for several reasons. Firstly, very little record keeping of exotic animals is legally mandated in most countries so little information is tracked. Secondly, many exotic pets are actually trafficked illegally and are, therefore, even more under the radar. The illegal trade in exotic pets, according to TRAFFIC (2011), a wildlife trade monitoring network, is worth billions of dollars a year. According to the American Society for the Prevention of Cruelty to Animals (ASPCA, 2011), exotic animals end up with their owners in one of three ways: extracted from their native habitat, bred in captivity, and sold off as surplus from zoos, circuses, and the like. Exotics are bought and sold mainly through the Internet, specialty magazines such as *Animal Finder's Guide*, *Animal Marketplace*, and *Animals Exotic and Small*, and local exotic animal auctions.

Not only does our consumption of pets and pet products bring a wide variety of activity, but as animal geographers we can also think about where and how animals are used to market products to us as humans generally, not just pet caretakers. Travis the chimp had appeared in ads for Coca-Cola and Old Navy. Advertisers are trying to catch people's attention as consumers, and animals offer just the ticket. Ritu Esbjörn (2007) highlights three ways businesses utilize animals: by using the animals themselves to sell a product, by using animals as symbols for something else, and increasingly by using animals to equate concern for the environment with the product being sold. In the first case, think of how animals are used to sell products like car insurance (the Geico gecko) or cereal (Tony the Tiger and Frosted Flakes). Often heavily anthropomorphized, the animals in these types of ads are trying to entice you to make purchases because you like the animals being presented. The next advertisement might be trying to convince you of the freedom and power you'll feel buying a certain vehicle because the ad shows it driving along next to galloping horses. Indeed, the notion of horsepower helped society transition from an animal-based transportation system to a machine-centered one. Horsepower refers to the power or weight of a horse pulling something and was originally used to compare steam engines to horses; it is used today to refer to the power of gas engines. The very next ad might use butterflies and singing birds to sell you on the idea of a fuel-efficient car that, if you bought it, would enable you to help the environment and still get around. Using animals to appeal to our sense of humor can also help promote a product. Thus animals are often put into unnatural situations to get a laugh and make you remember the brand the next time you're at the store. Advertising can also be used to reinforce cultural ideas about "good" or "bad" animals and where animals should be—their proper place. Consider commercials for pest control that construct other species as (1) pests and "bad" and (2) out of place in one's home or near one's body. The entire industry depends upon keeping people afraid of ants, termites, cockroaches, mosquitoes, spiders, opossums, raccoons, and so on. This is not to say that these beings have no negative impact for humans or their homes, but that, in a geographic sense, these animals must be kept in their place, away from humans, while "good" animals like cats and dogs actually make a home a home. Economically, then, multiple animal geographies are at work, from the commodity chain of pets (both legal and exotic) to the use of animals in advertising; therefore, our love of all things animal ends up being good for capitalism as well.

THE CULTURAL LANDSCAPE

As we discussed in chapter 1, the cultural landscape is the built environment—all the things that humans create from a small scale to a large scale. When we apply this concept to pets and the role of animals in culture, we want to think about where and how we encounter animal others and through which cultural mediums. We will begin with those animals closest to us—our pets—to see what animal geographers have learned about the place of pets in the home. Rebekah Fox writes that "pets occupy a liminal position on the boundaries between 'human' and 'animal,' appreciated by their owners as 'minded individuals' or friends, capable of rational thought and emotion, yet also treated as objects or possessions to be discarded if they do not conform to human expectations and values" (2006, 526). Pets are seen, then, as hybrid beings—a combination of nature and culture. Many people often differentiate a pet from a companion animal to demarcate how they view their nonhuman. A companion animal is a pet, but the attitude of the human toward the animal is not one of ownership or seeing the animal as property—an association they feel the word *pet* connotes. In Fox's research she found that people rely on popular pet psychology via books, television shows, magazines, and even veterinarians to understand their pets' behaviors. Whether dogs need to be part of a pack with a strong leader or cats need to go outside the home because it's their natural instinct, how we relate to the animals in our homes quite often depends on what other sources of information tell us; thus the human-pet experience is heavily mediated. The experience, however, also depends equally on the intimate day-to-day existence within the home. Fox found that in actual practice most humans anthropomorphized their animals and claimed they understood their feelings or why they were behaving in certain ways (either positively or negatively). This breaking down of the human-animal divide is a key aspect of pet keeping as humans and animals become "kept" by each other—the humans responding to the animals and vice versa. Thus, the success or failure of pets in particular homes depends upon those specific situations and not on the breed or species or humans as uniform groups. What Fox asks us to keep in mind, in addition, is that not all pets have such interactions. Her study focused on more traditional pets such as dogs and cats, but these animals are still seen as "closer" to humans than animals such as reptiles, spiders, or fish that are also kept by people.

In another study exploring the ways in which dogs become family, Emma Power (2008) concludes that they do so in three ways. Dogs are constructed as "furry children" to be cared for and watched over closely, as pack animals where the human needs to be the pack leader, or as beings with their own individual agency and subjectivity. Specific actions that fall into these categories include things like worrying over eating habits (furry children), controlling access to food or places within the home (being pack leader), and responding to a dog's individual actions like suddenly bringing a toy or learning to play the dog's game (recognizing agency). While these categories are not mutually exclusive, she argues that the "more-than-human family is shown to be a tenuous and contingent relation that is made, negotiated and sometimes falls apart as a result of the interaction that takes place between people and the particular animals that they live with" (536). Humans and their pets negotiate their relationship together; it is no longer understood as a one-way street with humans as the only ones who can act or control. What is so significant about a geographic analysis of humans and their pets interacting in their homes is that it reveals "a willingness to not just recognize, but to engage with and incorporate non-human others into family, and to explore other ways of 'being' within family" (546). Home as place is being transformed from a purely human arena to a complex intersection of multiple species.

Moving outward in space from the home, animal geographers have explored how pets are or are not part of public spaces. In a study on feral cats (cats that do not have human homes and have reverted to semiwild behaviors) in Hull in the United Kingdom, Huw Griffiths, Ingrid Poulter, and David Sibley (2000) show how these cats occupy multiple unique intersections of identities as they live very public lives. The authors see feral cats as occupying an elusive zone somewhere along the domestic-wild spectrum and note that "those animals which transgress the boundary between civilization and nature, or between public and private, which do not stay in their allotted space, are commonly sources of abjection, engendering feelings of discomfort or even nausea which we try to distance from the self, the group, and associated spaces (but which we can never banish from the psyche)" (60). Furthermore, "those people who are close to ferals or who try to engage with ferals, particularly the feeders, may themselves be viewed as discrepant by both council officers [local politicians] and neighbors" (65). Visibly we see feral cat colonies usually in abandoned or less-busy urban or suburban areas. The cats may congregate to sleep and interact with one another. Humans may support these colonies by feeding them, thus encouraging them to stay in one area, trapping them for veterinary care (medicinal or spay/neuter) and ear notching (which marks them as feral and "fixed") and releasing them back to the same area. The cats can largely fend for themselves but can become a nuisance to nearby humans because of noise, territorial marking, and digging, or just by the fact that they are there. Feral cats are one result of the human-animal relationship as pets move out from the confined space of the home and, in a way similar to the dog stealing, upend social norms about which animals should be cared for and how. Whether or not feral cats are in the "right place" ends up being about local human cultural and political geographies.

The issue of pet animals being in the "right" or "wrong" place can also be seen in the increasingly visible presence of off-leash dog parks especially in the United States. Under most urban and even suburban conditions, ordinances legally require dogs to be on leashes when outside of the home, which has made it difficult for human owners to find locations where their dogs can run free, socialize, and expend pent-up energy. In a case study of the politics of dog parks, Julie Urbanik and Mary Morgan (2012) highlight how dog parks are embroiled in larger conflicts over the "place" of dogs. The very act of trying to make a public place for animals runs counter to a long tradition of attempting to keep humans and animals in separate spheres—whether those animals are wild or domestic. Many people agree that some type of public spaces—national parks or wildlife refuges—can be places designed for animals, but that more local public parks should be for people only. If people want to have dogs, they need to care for their dogs in their homes and not in public areas where dogs can be off-leash and potentially out of control; the wild exuberance of dogs (barking, running, jumping, marking) is exactly what makes many people feel threatened along with creating a sense of being invaded by another species. On the other hand, those with dogs argue the animals need exercise to help keep everyone safe and part of the family, and giving them a place to play does not differ from designating areas of parks specifically for children, tennis games, Frisbee golf, or any other activity. Dog park proponents are actually making quite radical claims on living in a transspecies community (Wolch, West, and Gaines 1995). As dogs become more and more ubiquitous in our private lives, the rise of dog parks (over two thousand in the United States alone) on the landscape highlights how this relationship is moving into the public realm. We can see how this particular case echoes the work by Howell on pet cemeteries, both acts of transgression challenging the human-animal divide.

Pivoting to an entirely different aspect of the cultural pet landscape, we can turn to the problem of pet overpopulation. With millions of companion animals being killed every year because they do not have homes, private and public shelters have been working very hard to find ways to reduce this emotionally and financially expensive byproduct of modern-day pet keeping. Shelters have been pushing the spay/neuter campaign, reasoning that if people would get their dogs and cats spayed or neutered, society could drastically cut down on the number of animals being needlessly killed and warehoused. Julie Urbanik (2009) explores the ways in which gender comes into play with respect to who will get their animals "fixed" and where. As it turns out, in the United States, men seem to be those most concerned with neutering their animals, and a group in Utah had the bright idea to target men by hosting spay/neuter events at Hooters restaurants. These Hooters for Neuters campaigns have gone on around the country for several years—sometimes as simple fund-raisers, sometimes as on-site mobile spay/neuter events. While most people would not think of spay/neuter campaigns as problematic or controversial, this one has gotten people's attention because of the place—Hooters. Some people, both elected officials and feminist groups, argued that this particular campaign reinforced sexism by relying on women as sexual objects to "sell" the need to spay/neuter pets. Using one form of domination to "fix" another is seen as inappropriate: in essence, Hooters was the wrong *place* to conduct spay/neuter campaigns. Other people argued that it was the perfect *place* because it is a location where men are already comfortable being, therefore removing the "emasculating" stigma of having pets neutered.

David Lulka (2009) asks us to consider the American Kennel Club (AKC) and the ways in which it is a *place* where the ideal dog is constructed. Currently more than four hundred dog breeds exist in the world, 157 of them listed with the AKC, the largest registry in the world. The AKC was established in 1884 and today has a revenue stream of around $73 million each year from fees, shows, and licensing. He is trying to show the AKC's "ability to partially shape the corporeal character of dogs and the spatial distribution of dogs within the American landscape through its political activities" (535). In discussing which legislation the AKC supports and how members classify dogs themselves he shows that these endeavors "produce contradictory outcomes: the passion and love for dog breeds is countered by hierarchical modes of management, public concerns are increasingly incompatible with proprietary interests, [and] spatial formations extend inward and outward" (546). Although the AKC attempts to be the place where dogs are idealized, in practice the organization is incapable of doing so because it must contend with other groups (animal rights groups, dogfighters, governments) that have different visions of the ideal dog. So where and how specific dog breeds become normalized is actually a diffuse phenomenon.

Animal geographers have only tapped into a few aspects of the cultural animal landscape, so we are going to briefly explore two other arenas in which we can see human-animal relations being created. In the first case we can turn to human sports (animal sports will be covered in the next chapter). Human sports are about entertainment, excitement, skill, and competition. We love our team or players and feel solidarity with a particular place (whether on the local or national scale) as a fan. One of the main mediums for solidarity has been the use of mascots—representative stand-ins for a team.

> The main purpose of a mascot is to portray something that the whole team can
> identify with, that a whole stadium of fans can rally behind, and that can also

negatively affect the morale of the opposing team. For example, irrespective of whether it is biologically true or not, cats (Felidae) are stereotypically seen as being cunning, quick, and fierce. Perhaps this is the reason that this family of mammals is chosen so often to represent a sports team. Similarly, no sound-minded person would want to confront a ferocious bear. This persona of fear is exactly why such a mascot is chosen. (Garnett and Whiteman 2007, 1317)

Indeed, the mascot for the team of Michael Vick at the time of his arrest was the falcon—a bird characterized by its speed, eyesight, and deadly talons. Yet in other cases, like the American football team the Miami Dolphins, the mascot is chosen not to be ferocious but to represent a particular place (see table 3.1).

A second area with a lot of promise for animal geography is the use of animals in cultural visual media such as film and television. The prevalence of the visual in today's society means that cultural messages about animals often come to us this way; therefore the feedback loop between what we see and know about nonhumans is shaped by the ways in which animals are presented to us, which in turn shapes what we want to see of them. Films and television as cultural artifacts "[intensify] already existing cultural responses to animals, whether these have to do with, say, sentimentality, brutality, aesthetics, or fascination" (Burt 2007, 1205). Visual media about animals

Table 3.1.   Animal Mascots of Select US Major League Sports Teams

Bears	**MLB**: Minnesota Twins (*T.C. Bear*), **NBA**: Houston Rockets (*Clutch*), Utah Jazz (*Jazz Bear*), Memphis Grizzlies (*Grizz*), **NFL**: Chicago Bears (*Staley Da Bear*), **NHL**: Boston Bruins (*Blades*), St. Louis Blues (*Louie*), Minnesota Wild (*Nordy*)
Birds	**MLB**: Baltimore Orioles (*The Bird*), St. Louis Cardinals (*Fredbird*), Pittsburgh Pirates (*Pirate Parrot*), Washington Nationals (*Screech*), **NBA**: Atlanta Hawks (*Harry the Hawk*), **NFL**: Baltimore Ravens (*Poe, Rise, Conquer*), Arizona Cardinals (*Big Red*), Atlanta Falcons (*Freddie Falcon*), Philadelphia Eagles (*Swoop*), Seattle Seahawks (*Blitz*), **NHL**: Pittsburgh Penguins (*Iceburgh*), Washington Capitals (*Slapshot*), Atlanta Thrashers (*Thrash*), Chicago Blackhawks (*Tommy Hawk*), Anaheim Ducks (*Wild Wing*)
Cats	**MLB**: Arizona Diamondbacks (*Baxter the Bobcat*), Detroit Tigers (*Paws*), Kansas City Royals (*Sluggerrr*), **NBA**: Charlotte Bobcats (*Rufus Lynx*), Denver Nuggets (*Rocky the Mountain Lion*), Indiana Pacers (*Boomer*), Portland Trail Blazers (*Blaze the Trail Cat*), Sacramento Kings (*Slamson*), **NFL**: Cincinnati Bengals (*Who Dey*), Jacksonville Jaguars (*Jaxson de Ville*), Carolina Panthers (*Sir Purr*), Detroit Lions (*Roary*), **NHL**: Los Angeles Kings (*Bailey*), Florida Panthers (*Stanley C. Panther*)

Source: This table includes a selection of animal mascots from Major League Baseball (MLB), the National Basketball Association (NBA), the National Football League (NFL), and the National Hockey League (NHL).

include science-based shows that emphasize a detached scientific perspective, sentimental pieces that either allow us to empathize with nonhumans or delight in them as living beings, overtly political shows that describe the exploitation of animals by humans, and movies or television shows that construct animals as monsters. All of these categories are rooted in different human-animal geographies and ensuing messaging about how we are to view and experience these animals.

The global success of films like *Blue Planet*, *Winged Migration*, and *March of the Penguins* exemplifies a commercialized documentary format where a human narrator speaks about the animals, yet no humans are seen on the screen. While educational, these types of films can also serve to reinforce the separation between humans and animals—essentially by making humans the ones that are "out of place" in pristine environments. In sentimental films and shows like any Walt Disney movie, animals are heavily anthropomorphized—often wearing clothes, speaking, thinking, and acting like humans. This type of visual medium often creates porous boundaries between humans and animals: whether the animals are wild, domesticated, in captivity or not, the message is that animals are really like us. With politically motivated films like *Meet Your Meat* about industrial animal farming or *Lolita* about a wild-caught killer whale who has spent over forty years performing in a swimming pool, the aim is to expose the places in which specific human-animal practices occur and how. The aim is as much to reveal human-animal geographies as it is to decry a particular practice. Finally, movies about animals as monsters—from *Anaconda* to *Jaws* to *The Grey*—serve to do the opposite of anthropomorphizing animals.

> Animal monsters describe the animals as "others," as beings who are opposites of humans. They portray characteristics that are very unlike those portrayed by what are understood as stereotypical humans, and hence aggression overtakes morality, instincts overtake reason, bodily functions overtake purity, etc. However, animal monsters may also include animals that behave similarly to human villains (for instance, cold intelligence and sociopathy are evoked). Thus, animal monsters are constructed as opposites of "good" human beings. (Aaltola 2007, 1199)

The main problem with animal monsters is that they are out of place in a society that places humans at the top. When animals hunt humans (*Jaws*), have power over them (original *Planet of the Apes*), or become crossed with humans (*The Fly*), the result is terrifying for "us" because the animal monsters gain an agency and power that is not supposed to be. While these monsters may be challenging human-animal power relations, they can also challenge our emotions. We are terrified of animal monsters, but in many cases we also feel empathy for their existence especially when we have created them. Take for example *King Kong* and the apes in *Rise of the Planet of the Apes* because we see what humans have done with their power to capture and control. Films like these require us to do deeper reflection on just who is a "monster" if humans are doing things to animals that turn them into "monsters."

ETHICAL/POLITICAL GEOGRAPHIES

In this section, we will explore the intersection of ethics and politics through three topics: bestiality, pet-specific legislation, and animal activism. Remember, ethics is about notions of right and wrong behavior, and politics is the conflict over whose vision of society and right/wrong behavior will dominate. The legal arena—through case and legislative laws—is most often where animal conflicts are addressed and possibly resolved. So as animal geographers we can consider the legal geography of human-animal relations in terms of where specific practices are legal or not, but we can also get a glimpse into which places of human-animal interaction matter in terms of ethical/political conflicts.

Bestiality and zoophilia are uncomfortable subjects for most people as they are considered quite taboo. Bestiality is having sexual relations with an animal while zoophilia is more about having a loving, consensual sexual relationship with a specific animal. Zoosadism is when people derive sexual pleasure from inflicting pain on animals. For all three forms animals used include dogs, horses, snakes, donkeys, sheep, gerbils, and goats. While evidence suggests bestiality has always existed in some form in human cultures, it has nearly always been rejected in Judeo-Christian and Islamic cultures as a sin and moral transgression. However, some, like Peter Singer, author of *Animal Liberation,* argue that zoophilia may in fact be a sexual orientation akin to hetero- or homosexuality, and that we should approach the issue with today's knowledge of other species, not simply adhere to historical cultural or religious taboos (Singer 2001). The issue of laws being used to monitor boundaries between humans and animals has been taken up by David Delaney: "To the extent that modern understandings of what it is to be human are dependent on particular conceptions of nature, it is reasonable to suggest that legal discourse cannot be 'neutral' with respect to competing conceptions of nature" (2001, 488). With the case of bestiality the law has historically been used to reinforce a vision of what is "natural" in terms of human sexual relations and to create boundaries to prevent transgressions. He reminds us that "the physical realization of the views that nature is an exploitable resource and that animals are appropriate objects of experimentation is inseparable from the legal conception of property that confers on legal subjects [humans] the right to use and even destroy their property objects" (489).

For Michael Brown and Claire Rasmussen (2010) the case of a man dying from acute peritonitis from a perforated colon, caused by his being anally penetrated by a horse in Washington State, was a chance to explore not only the legal geographies of human-animal boundaries but also the more complicated ways in which the boundaries between humans and animals are policed. They document the five major public discourses that appeared in response to the publicity this event created: (1) worry that the state would become a magnet for bestiality because the state has no law against it (states, not the federal government, have jurisdiction here), (2) concern about the moral implications because of animals' inability to consent, (3) sexual abuse of the horse, (4) equation of animals with children, making the act akin to pedophilia, and (5) preservation of our humanity. They conclude that "the discourses attempt to condemn the act of being fucked by a horse not to protect either horses or even individuals but to express concern about the transgression of boundaries" (166). This boundary rupture between humans and animals via sexual acts is so threatening because it denies the rigidity of the lines between humans and animals in the first place and calls into question human superiority. "Ironically, both the pleasure of bestiality for the practitioner and the horror/ humor of the shocked observer rely upon [a] projection of humanity on the animal" (174). Furthermore, the consent argument contradicts our other uses of animals: if animals can't be considered able to consent to have sex then how can they consent to being killed for food or milked for milk or used for clothing or neutered or any one of the other numerous ways we make use of and manage animals.

With respect to the ethical/political slippage directly related to pets we can consider puppy mills and breed-specific legislation (BSL). Puppy mills, largely a problem within the United States, are basically puppy factories. While legal structures are in place for licensing breeders, historically these laws have rarely been enforced. This laxity has meant that breeders often keep the dogs in what activists consider to be inhumane conditions—a lifetime in wire cages, no exercise or access to the outdoors, no human affection, females constantly pregnant until they are physically spent, and little veterinary care. Animal advocacy groups such as the Humane Society of the United States have been successful in portraying dog breeding in the negative light of puppy mills and getting regulatory legislation passed. The general public likes dogs and the idea that one's beloved companion came from a situation where other dogs were being mistreated doesn't sit well. Proponents of more lax regulations on breeders argue that the dogs are their property, that they need to earn a living and make a profit, that they do give plenty of basic care to the animals, and that the public shouldn't be bullied by animal extremists or a few bad breeders. As a result of this conflict, thirty-three US states have laws regulating the most egregious puppy mill practices, but no law specifically bans puppy mills as a whole. Canada, Australia, and the United Kingdom also have similar regulations in place.

Specific breeds seem to cause a lot more concern and legal action. BSL is used to ban certain dogs from certain places. So on the one hand we want to regulate puppy mills, but we don't want certain dogs to be around. For example, the city of Denver, Colorado, banned pit bulls, as did several cities in Missouri and Kansas. The dog bans normally target larger, more menacing breeds like pit bulls, chows, or rottweilers. Arguments for BSL have to do with public safety and preventing dog bites and mauling, but also to keep "undesirable" humans like drug dealers or dogfighters away from towns/counties. Opponents of BSL argue that breed prejudice is unjustified because dogs are, firstly, individuals with a spectrum of temperaments, so claiming that all pit bulls are bred to be aggressive toward humans incorrectly generalizes. Secondly, opponents argue that dog behavior comes from humans, so the blame for problematic animals should be put on owners. In both of these cases, at stake is when and how animals—and human uses of them—are out of place. The legal and legislative systems provide one method of policing "proper" human-animal boundaries *and* right/wrong behavior toward animals.

What happens when conflicts over practices on animals become extreme? In this third instance of ethical/political conflict, we want to consider more extreme forms of animal activism—those that involve purposefully breaking the law to raise awareness of perceived injustices. Are these kinds of animal rights activists, often associated with nonhierarchical groups such as the Animal Liberation Front (ALF) and the Earth Liberation Front (ELF), terrorists or freedom fighters? The US government considers animal rights extremists terrorists because they seek to cause economic and emotional damage to legal practices with animals (Federal Bureau of Investigation [FBI] 2005). In fact, the Animal Enterprise Terrorism Act (AETA) codifies animal activism and makes the criminal disruption of any legitimate business that uses animals a felony (AETA 2011). Animal rights activists, however, see themselves more as freedom fighters who have to go to extreme measures to save animals because the government won't. In fact, they see the treatment of animals as so problematic that they will risk fines, jail time, and even their lives to end animal suffering. Furthermore, they argue that the perpetrators of violence to animals are the real terrorists because they target innocent animal lives. Legislation like AETA is seen as legitimating the profit motives of businesses over the rights of sentient beings not to suffer. Figure 3.1 shows self-reported

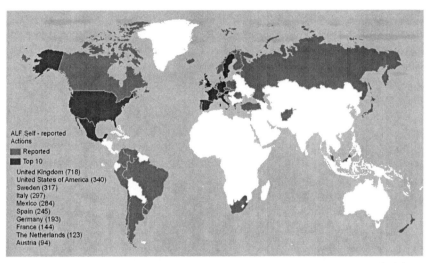

Figure 3.1. Total Number of ALF Self-Reported Attacks. *Source:* Data compiled from the ALF website Bite Back (http://www.directaction.info), covering 2003–2010.

ALF actions around the world. Does the top ten list of countries hold any surprises? These actions range from painting graffiti on storefronts to breaking and entering to destroying property or releasing animals. At this time no human has died in an ALF action.

The ethical/political issues regarding pet and cultural animal relations are obviously larger than these three examples, but these examples do serve as a solid mapping of the different locations in which contested practices take place—in legal systems, in legislative systems, and in public/private spaces of protest. At stake is who has the power to determine how animals can be used and where. In this chapter we have seen how an animal geography lens can be used to shed light on pets as well as larger cultural markers. *Who* you are in terms of your treatment of animals often depends on *where* you are—spatially and culturally. Whether you want to reinforce the human-animal divide or cross it sexually, create a more-than-human family with a dog or see the dog as a purely experimental tool or economic object, or utilize dog parks or ignore them, our identities, language, and everyday spaces are imbued with complex interspecies power geometries.

DISCUSSION QUESTIONS

1. Are there other types of pet relations or cultural markers that haven't been covered here? How would you bring those into an animal geography perspective?

2. Does Tuan's concept of dominance and affection apply to your relationship with any pet animals? Where/how did you learn how to interact with these animals? How do your different identities come into your pet relationships?
3. Do you agree with Nast that modern pet keeping is diverting attention from more pressing human concerns? What evidence do you have for your views?
4. In what ways do market forces appeal to emotions when it comes to pet keeping?
5. What is a more-than-human family? Where/how do you see it constituted in your own life (either with your own family or in others)? Who is keeping whom?
6. Can organizations such as the ALF really only be seen as either terrorists or freedom fighters or can they be categorized as something else entirely?
7. What other media can you think of that portray animals as humanlike, monsters, or some combination of human and animal? What are the messages about human-animal relations and the proper place of animals in these examples?

KEYWORDS/CONCEPTS

Animal Enterprise Terrorism Act (AETA)
bestiality
breed-specific legislation (BSL)
critical pet studies
dominance and affection
exotic pets (trade)
legal geography
more-than-human family
pet versus companion animal
puppy mills
zoophilia
zoosadism

PRACTICING ANIMAL GEOGRAPHY

1. Compare and contrast, from a geographic standpoint (using the geographer's toolkit from chapter 1) two different pet breeds.
2. Make a map of the pet-related places in your community. What does this map reveal about the human-pet bond in your area?
3. Go to your local grocery, liquor, or department store. What are animals (or their parts) selling you? Why and how?
4. Interview ten family members or friends about their attitudes toward the ALF and reflect on whether their attitudes come from firsthand experience or media. What might this say about the role of media in shaping people's perspectives about animal-related issues?

RESOURCES

American Enterprise Terrorism Act: http://www.govtrack.us/congress/bill.xpd? bill=s109-3880

American Kennel Club: http://www.akc.org

Animal Liberation Front: http://www.animalliberationfront.com

Behind the Mask (film about animal rights activists): http://www.uncagedfilms.com/ behindthemask.php

Disney films: http://disney.go.com/movies/index

Madonna of the Mills (film about puppy mills): http://madonnaofthemills.com

The Tiger Next Door (film about exotic cats in captivity): http://thetigernextdoor.com

TRAFFIC: http://www.traffic.org

REFERENCES

Aaltola, Elisa. 2007. "Animal Monsters in Film." In *Encyclopedia of Human-Animal Relationships: A Global Exploration of Our Connections with Animals*, edited by Mark Bekoff, 1197–1203. Westport, CT: Greenwood.

American Pet Products Association. 2011. "National Pet Owners Survey 2010–2011." Accessed February 2. http://www.americanpetproducts.org/press_industry trends.asp.

American Society for the Prevention of Cruelty to Animals. 2011. "Exotic Pet Trade." Accessed March 2. http://www.aspca.org/fight-animal-cruelty/exotic-pet-faq.aspx.

Animal Enterprise Terrorism Act. 2011. Accessed June 19. http://www.govtrack.us/ congress/billtext.xpd?bill=s109-3880.

Brown, Michael, and Claire Rasmussen. 2010. "Bestiality and the Queering of the Human Animal." *Environment and Planning D: Society and Space* 28:158–177.

Burt, Jonathan. 2007. "Media and Film." In *Encyclopedia of Human-Animal Relationships: A Global Exploration of Our Connections with Animals*, edited by Mark Bekoff, 1203–1207. Westport, CT: Greenwood.

Delaney, David. 2001. "Making Nature/Marking Humans: Law as a Site of (Cultural) Production." *Annals of the Association of American Geographers* 91 (3): 487–503.

Esbjörn, Ritu. 2007. "Advertising and Animals." In *Encyclopedia of Human-Animal Relationships: A Global Exploration of Our Connections with Animals*, edited by Mark Bekoff, 1194–1197. Westport, CT: Greenwood.

Federal Bureau of Investigation. 2005. *Terrorism 2002–2005*. Washington, DC: Government Printing Office.

Fox, Rebekah. 2006. "Animal Behaviours, Post-human Lives: Everyday Negotiations of the Animal-Human Divide in Pet-Keeping." *Social and Cultural Geography* 7 (4): 525–537.

Garnett, Susan, and Howard Whiteman. 2007. "Sport and Animals." In *Encyclopedia of Human-Animal Relationships: A Global Exploration of Our Connections with Animals*, edited by Mark Bekoff, 1315–1317. Westport, CT: Greenwood.

Griffiths, Huw, Ingrid Poulter, and David Sibley. 2000. "Feral Cats in the City." In *Animal Spaces, Beastly Places: New Geographies of Human-Animal Relations*, edited by Chris Philo and Chris Wilbert, 56–70. New York: Routledge.

Howell, Philip. 1998. "Flush and the Banditti: Dog-Stealing in Victorian London." In *Animal Spaces, Beastly Places: New Geographies of Human-Animal Relations*, edited by Chris Philo and Chris Wilbert, 35–55. New York: Routledge.

———. 2002. "A Place for the Animal Dead: Pets, Pet Cemeteries and Animal Ethics in Late Victorian Britain." *Ethics, Place and Environment* 5 (1): 5–22.

Lulka, David. 2009. "Form and Formlessness: The Spatiocorporeal Politics of the American Kennel Club." *Environment and Planning D: Society and Space* 27:531–553.

Nast, Heidi. 2006a. "Critical Pet Studies?" *Antipode* 38 (5): 894–906.

———. 2006b. "Loving . . . . Whatever: Alienation, Neoliberalism and Pet-Love in the Twenty-First Century." *ACME: An International E-Journal for Critical Geographies* 5 (2): 300–327.

Power, Emma. 2008. "Furry Families: Making a Human-Dog Family through Home." *Social and Cultural Geography* 9 (5): 535–555.

Singer, Peter. 2001. "Heavy Petting." Accessed March 7, 2011. http://www.nerve.com/Opinions/Singer/heavypetting.

TRAFFIC. 2011. "Wildlife Trade: What Is It?" Accessed March 12. http://www.traffic.org/trade.

Tuan, Yi-Fu. 1984. *Dominance and Affection: The Making of Pets*. New Haven, CT: Yale University Press.

Urbanik, Julie. 2009. "Hooters for Neuters: Sexist or Transgressive Animal Advocacy Campaign?" *Humanimalia* 1 (1): 41–67.

Urbanik, Julie, and Mary Morgan. 2012. "A Tale of Tails: The Place of Dog Parks in the Urban Imaginary." Unpublished manuscript.

Wolch, Jennifer, Kathleen West, and Thomas Gaines. 1995. "Transspecies Urban Theory." *Environment and Planning D: Society and Space* 13 (6): 735–760.

Chapter Four

Beasts of Burden:
Geographies of Working Animals

On May 1, 2011, a Navy Seal team that included a dog entered infamous terrorist Osama bin Laden's compound in Pakistan and killed the al Qaeda leader. The military operation came at the culmination of a nearly decade-long manhunt to get the mastermind behind the 9/11 attacks in the United States. While the identities of this elite military unit have remained anonymous, the use of the dog has stirred people's imaginations. The dog, probably a German shepherd or a Malinois, was trained to jump out of aircraft and provide any number of support services to the Seal unit. An estimated twenty-seven hundred dogs are serving in the US military today and that number is on the rise (Bumiller 2011). Dogs are loyal to humans, have incredible sensory capabilities through their nose and ears, and can move in ways humans cannot; as such, they have become an increasingly important tool for everything from bomb detection to troop morale (Bumiller 2011).

Other dogs have also become famous for their role in major political events. For example, Laika, a mixed-breed street stray from Moscow, was the first dog launched into space in *Sputnik 2* in 1957. While she was hooked up to life support systems, *Sputnik 2* was not designed for recovery, and after her air supply and cooling systems malfunctioned, she likely passed away from overheating and stress hours after launch and long before the spacecraft fell back into the atmosphere and burned up in 1958 (Space Today 2011). Her death helped the Soviet Union beat the United States during the Cold War space race. Today she is memorialized with a statue and plaque near Moscow. Toto, the dog from *The Wizard of Oz* film, was played by a two-year-old female cairn terrier named Terry. She earned $125 per week—more than any of the munchkins in the film. She died nine years later after starring in twelve movies. What do Toto, Laika, and the Navy Seal dog have in common? Their celebrity is certainly a commonality, but their status as laborers is what interests us here. The focus of this chapter is on the spaces and places in which humans utilize nonhumans as workers. As we will see, this covers a wide range of activities, species/breeds, methods, and locations, and we will use our geographic toolkit to compare and contrast this set of human-animal relations. As with pets and broader cultural categories, working animals are all around us on a daily basis, whether we see them and know it, or not.

Working animals can be found in myriad human-animal relations. For our purposes we will sort these animals into three major categories: educational animals, entertainment animals, and service animals. The category of educational animals includes animals used for instruction activities such as dissection, anatomy, surgery, and laboratory classes; animals used in medicinal, military, academic, industrial, agricultural, or veterinary research; and captive animals in zoos with a conservation/education mission. In all three of these cases animals are used as educational tools to further human understanding of either animals, products, human systems, or the environment. Entertainment animals are those that are made to perform for the pleasure of humans. This category includes a wide range of activities from racing (e.g., dogs, horses), to circuses, captive shows (e.g., Seaworld), street performers (e.g., dancing bears, organ grinder monkeys), petting zoos, rodeos/eventing, fights, and TV/film work. Lastly, the service category includes draft animals (whether for agriculture or transport), hunting/herding dogs, assistance/therapy animals, and even logging elephants along with animals in law enforcement and the military. Figure 4.1 provides a visual graphic to map out these categories. As we can see, subcategories can overlap. For example, zoos are supposed to be about education, but they are also entertainment. Also, all of these animals are technically performing a service for humans; however, the purpose and places of these services differ greatly.

In order to make sense of this umbrella category of working animals, this chapter is going to provide a very broad overview of the three subcategories. Justice cannot be done to the subcategories in just one chapter partially because animal geographers have yet to fully venture into this area, but also because human–working-animal configurations exist in such sheer numbers. Additional resources for going more in depth in a particular area can be found at the end of the chapter. We will follow the same chapter subsections as in chapter 3, but we will focus on just one or two examples in each section.

HISTORICAL GEOGRAPHIES

Most types of working animals have a long history; however, determining what geographers refer to as a cultural hearth, or place of origin, is often

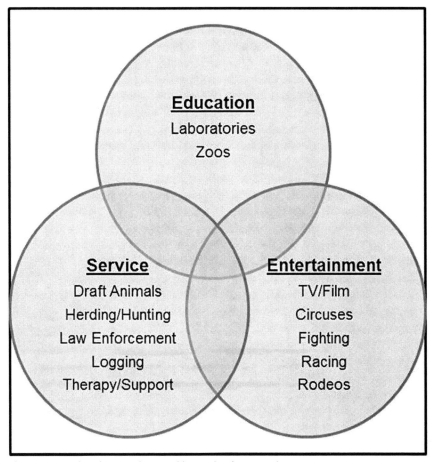

Education
Laboratories
Zoos

Service
Draft Animals
Herding/Hunting
Law Enforcement
Logging
Therapy/Support

Entertainment
TV/Film
Circuses
Fighting
Racing
Rodeos

Figure 4.1. The Three Major Working Animal Categories.

difficult. Practices were often independently invented, which means that the evidence we have can point to multiple origin locations. Furthermore, the evidence can be incomplete, meaning that while a practice may seem to have developed later in one location than another, nothing can be concluded because many cultures neglected to leave detailed explanations with full diagrams for future societies like ours!

With the subcategory of educational animals via captivity and zoos we are looking at a complex history with evidence going back several thousand years and spanning Africa, Asia, Europe, and more recently Central and South

America. Throughout the history of empires, animals have been used to demonstrate power and control over territory as well as for diplomacy and gift giving between powers. We have evidence for what is possibly the world's first zoo in southern Egypt. In the ancient city of Hierakonpolis ruins of a zoo from about 3500 BCE have been unearthed, revealing a collection of animals like baboons, elephants, and hippos. The animals were buried in a section of the city where royalty was buried—demonstrating their valued status for those in power (Rose 2010). During the Middle Ages, British royalty collected exotic animals from afar and kept them in "menageries" in small cages or pits where curious people could view them. King Henry I (1068–1135) established the "Royal Menagerie" in Woodstock, which was later moved to the Tower of London. For hundreds of years wild animals were captured or given as gifts from other dignitaries (Nichols 1999). Gifts to British royalty included African elephants, leopards, lions, camels, a polar bear, and a porcupine. In 1519, Hernán Cortés described a zoo so large in Mexico that three hundred people were needed to care for the animals (Hancocks 2001).

What we recognize as the modern zoo began in Europe with the Vienna Zoo in 1752 and the Paris Zoo in 1793 (Walker 2001). While these zoos were still part of a colonial power structure—most of the exotic animals came from areas that were in the process of being colonized or newly explored—collectors began to develop classification systems and gather baseline information about the animals. The main purpose of these zoos was to provide a way to see and study the different species of the world. Very little attention was paid at all to the lives of the animals in captivity, and most cages were bare and empty in order to better see the animals. Over the course of the twentieth century, zoos have evolved to be places of education—both for the general public and for biologists and veterinarians. The emphasis today is on what we might consider educational entertainment where, yes, the animals are kept in cages (or euphemistically habitats or enclosures) but the goal is to get people to learn about the species living in the world and the problems they face (Norton et al. 1995).

The use of animals for research purposes also dates back quite some time, but the cultural hearth seems to be in Europe. Franklin Loew and Bennett Cohen (2002) provide the following examples of the evolution of this practice. The first recorded example of animal experimentation comes from Aristotle, who used dissection to demonstrate differences between species. During the time of the Roman Empire, the Catholic Church banned human autopsies, so the physician Galen began doing dissections and autopsies on animals—both alive and dead. He is often considered the "father of vivisection." *Vivisection* is Latin for "live cutting," but animal activists often use the word today to refer to all animal research and experimentation. Middle Eastern physicians such as Avenzoar and Ibn al-Nafis in the thirteenth century provided information about the circulatory system by dissecting animals, but the modern use of animals for research took off in Europe. Loew and Cohen argue this jump forward was caused by the confluence of the Scientific Revolution, which inspired a determined interest in understanding the world, the Age of Exploration, which provided an increasingly vast world to understand, and medicine, which made great advances. The vast majority of this type of research was either pure research—simply to understand how life-forms work—or applied research—research with the purpose of helping humans. Early modern examples of applied research would be Louis Pasteur's discovery of germs by giving anthrax to sheep and Emil von Behring's use of guinea pigs to isolate the diphtheria bacteria and develop a vaccination for the disease, winning him a Nobel Prize in 1901. A related connection to Behring's work is the Iditarod dogsled race in Alaska. This race is run in part to commemorate the dog relay teams that successfully ran 674 miles from Nenana to Nome in 1925 to bring vaccinations to Eskimo children under threat from an outbreak of diphtheria (AlaskaNet 1995).

Today animal research covers a wide range of activities and goes on in nearly all countries around the world. In addition to pure and applied research, animals are used in toxicity testing for chemicals, cosmetics, and drugs, by the military for weapons testing and training, and in all levels of education. The most ubiquitous animal research model is the mouse. Why mice? Before lab mice were lab mice they were novelty pets. Breeding "fancy mice" (mice with a variety of coat colors or specific traits) became popular in the United States around the turn of the twentieth century after they were imported as a hobby from Victorian England.

> In 1930s England, mouse breeders could cash in on the demand for full-length women's coats made of mouse skins, which took 400 skins and sold for $350. These mouse breeders and fanciers essentially routinized the activity of mouse breeding in captivity . . . and thus provided scientists with both a unique mammalian material resource and a broader practical context in which mouse breeding was an accepted cultural activity. (Rader 2004, 32)

In 1902 Harvard geneticist William Castle bought some fancy mice from Abbie Lathrop, a mouse breeder in Granby, Massachusetts (Rader 2004). It was Castle's student, Clarence Little, who, in 1909, successfully stabilized the first inbred mouse strain that could be used for biological research. In 1929, the nonprofit Jackson Labs was founded in Bar Harbor, Maine, and has since served as one of the main producers of laboratory mice for research in the United States. Mice can be cared for inexpensively, handled easily, and manipulated well genetically as well as reproducing rapidly.

Indeed, the use of mice in developing technologies like cloning, genetic modification, and genetic engineering (adding or removing genetic material) has pushed animal research to a new level. In 1998, the US Patent and Trademark Office (USPTO) issued a controversial patent for the first mammal, a line of genetically modified laboratory mice called Oncomouse (designed for susceptibility to cancer) to Harvard University (USPTO 1998). Hundreds of living beings have now been patented. On July 5, 1996, Dolly the sheep was born in Scotland, the first mammal to be cloned from an adult cell (Kolata 1998). On January 8, 2001, Noah, a cloned gaur cow, was born in Massachusetts (Advanced Cell Technology [ACT] 2001). The list goes on, and in the last couple of decades, the role of genetic technologies in changing laboratory animals has rapidly advanced and continues to do so. According to Jody Emel and Julie Urbanik (2005), four major categories of commercial activity have developed in this area: (1) producing biological and medical goods, such as genetically engineered pigs for use in xenotransplantation (cross-species organ transfers); (2) pharming, or using female animals as bioreactors able to excrete material through their milk; (3) increasing agricultural output; and (4) satisfying cultural desires such as cloning extinct species or beloved pets. Animal research then, like keeping animals captive, has a long history that has developed through stages into accepted industries in today's society.

With the subcategory of entertainment animals we can also trace many of the histories back to the practices of ancient empires. For cultural hearth areas of circuses we can turn to Egypt, Greece, and China (International Circus Hall of Fame [ICHOF] 2011). The word *circus* may be traced to the Circus Maximus in the Roman Empire—a stadium that could hold 250,000 people. Spectators enjoyed a variety of animal events from racing to tricks to fighting. However, some archeologists claim that the Roman circus was based on similar events that had been held in Egypt and Greece. In 108 BCE Emperor Wu of Han in China held a circus for foreign guests, the first such reported event, and while Western texts referred to it as a circus, the Chinese called it an acrobatic performance, but it did, apparently, include animals (ICHOF 2011). What we think of today as the traveling, three-ring circus developed with Philip Astley and John Hughes in 1768 in England (PBS 2011). Astley was the first to bring horses to an indoor arena where they did tricks while Hughes, a rival of Astley's, came up with the name *circus* for his own indoor show. By the late 1700s entrepreneurs began taking menageries on the road, charging spectators to see the exotic animals. British equestrian, John Bill Ricketts, put on the first circus in America in 1793 in Philadelphia. Joshuah Purdy Brown's circus was the first to use the tent format in 1825, and American Isaac van Amburgh was the first to put his head inside a lion's mouth in 1833. Railroads began to transport circuses in 1832. In 1851 George Bailey added a menagerie, including elephants, which quickly became the stars of the circus. Jumbo, the first famous circus elephant, was caught in the wild as a baby in southern Sudan and lived in the Paris and London zoos before being sold to the Barnum and Bailey Circus for $10,000 in 1882 (Chambers 2009). Jumbo died in 1885 after being hit by a train in Canada. Circuses were so popular in the early twentieth century that Lenin nationalized the circuses in the Soviet Union as part of celebrating "the people's art." Today, the Barnum and Bailey Circus still travels and performs with animals, but the rise of non-animal circuses like Cirque du Soleil and the controversy of using animals in circuses overall have tempered public support of this once beloved form of entertainment.

Horse racing as a sport probably emerged in central Asia where scholars believe the horse was first domesticated around five thousand years ago. Records across Asia and Europe document horse racing as an established sport that included both racing of horses with riders and racing of chariots with the horses pulling the chariots. The origins of modern horse racing with its focus on breeding occurred in England at the end of the crusades in the 1200s (Animal Planet 2011). Crusaders brought back Arabian horses to breed with local horses, and racing became a popular pastime of the upper classes. By the early 1700s horse racing was becoming more organized and professional with tracks being built around the country. While horse racing came to America with the British, it did not really get going until after the Civil War when it began flourishing (WinningPonies 2011). As a modern sport, it has had its ups and downs mainly from controversies surrounding gambling, organized crime, and the treatment of the animals.

Related to horse racing is the rodeo. A distinct sport that emerged in North America with the practices of Spanish ranchers and their workers called vaqueros. As they helped the ranch owners, the vaqueros handled a wide variety of livestock-related tasks: horsemanship, lassoing, branding, breaking wild horses, and so on. At some point very localized competitions began to take place when vaqueros finished cattle drives and had time on their hands. As the United States took land from the Spanish, American-style ranching emerged and these ranch hands also enjoyed competing and showing off their skills. The first rodeos were held in Arizona, Colorado, and Wyoming (Applebome 1989). Today, with advances in technology making the traditional lives of ranch hands much less necessary in the United States, the sport has continued in a professional context with crowds of people thrilled to watch both men and women test their skills.

Gambling and organized crime connections are more prevalent in animal fighting—especially dogs and roosters. Hanna Gibson (2005) provides a brief history of this sport. As a historic pastime, it was associated with many different cultures in Asia and Europe. Dogs fought animals such as bears and bulls in the Roman Colosseum. By the thirteenth century, dogs were pitted against bulls in England. Under Queen Elizabeth I (1533–1603), most towns had rings featuring the popular spectacle, upon which people gambled. All classes, including the queen, watched the sport from tiered benches. The bull was tethered by a rope connected to an iron stake while trained bulldogs took turns trying to latch onto the bull's nose, its most tender part. It usually lasted about an hour with the winner either the bull, who withstood the onslaught, or the dogs, who "pinned" the bull and pulled it to the ground. Bears were sometimes the stars of this sport in which a blinded bear was whipped by humans and bitten by bulldogs. Finally, the Humane Act of 1835 outlawed dogfighting and bull/bear baiting in England after growing discomfort with the practice became widespread. While bull/bear fighting never caught on in the United States, the Staffordshire bull terrier was brought to America in 1817 and dogfighting became a popular sport. Today dogfighting is illegal in both England and the United States; however, as an underground industry it remains popular. Cockfighting also has roots in Greece and Rome, but also in Southeast Asia, where evidence suggests humans first domesticated chickens. Cockfighting, like dogfighting, involves putting roosters in enclosed areas and inducing them to fight with each other. Banned in many countries, cockfighting continues to be a popular sport in Latin America and Asia.

Animals in television and film are, obviously, a modern invention. Animal training as a job category emerged with the expansion of the circuses, but the arrival of moving pictures opened up a whole new world for working animals. Jungleland, a Hollywood facility that opened in 1926 and was run by Louis Goebel, was home to such celebrity animal entertainers as Mr. Ed the talking horse and Leo the roaring MGM lion (Stagecoach Inn Museum 2011). Trained to do all sorts of stunts, wear human clothing, or just mug for the camera, animal celebrities have made us laugh, made us cry, and enticed us to buy their products for years, but they are not the only entertainers out there. In southern Asia, dancing bears have been a popular form of entertainment for hundreds—perhaps thousands—of years. Sloth bears are either bred or wild-caught and trained in India by the nomadic Kalandar people. In order to train the bear a hot metal rod is pierced through its nose (without anesthesia) and a rope inserted through the hole. The owner then trains the bear to stand, sway, and dance by manipulating the rope. Dancing bears have also been popular in Pakistan, Turkey, and even Bulgaria and parts of Eastern Europe. The United States enjoys its share of entertainment bears as well. Bear shows, such as those featured at Clark's Trading Post in New Hampshire, portray bears as happy performers (Clark's Trading Post 2011). The company's website states that the bear shows are "truly a tradition among White Mountain vacationers" and that "this entertaining and educational half-hour show is a thrill for all." The bears are unmuzzled and unleashed and perform tricks like riding Segways, jumping barrels, and posing. Entertainment animals certainly perform a wide spectrum of activities; our last subcategory—service animals—is nearly as disparate.

The global use of draft animals makes up the largest subcategory of all the working animals. Draft animals are those that are used for agriculture (pulling plows), pumping water, and transporting people and goods. The earliest direct evidence we have for draft animals comes from pictorial carvings showing oxen with plows in the Mediterranean area, Egypt specifically, from around six thousand years ago. The donkey was domesticated from the wild ass in Egypt sometime during this period as well. We also have llamas from South America, camels in the Middle East and Asia, water buffalo from south Asia, and goats from southwest Asia all coming into the folds of human societies as working animals between 8000 and 4000 BCE.

Elephants have also been used as draft animals. Elephant logging is primarily utilized in south and southeast Asia (Jayasekera and Atapattu 1995). Originally elephants were used as war animals, but as those jobs disappeared, they became strictly labor animals for the logging industry. The mahout, or elephant handler, puts a lot of time and effort into raising a working elephant. The mahout's key role is managing the elephant, which carries many responsibilities and learned skills. After intensive training and relationship building with the mahout, at twenty to twenty-five years old the elephant is put to work, and it will spend about thirty-five to forty years laboring. At around sixty-five the elephant is retired, and recently a retirement community has been developed for these animals in the Ngao district of Lampang, Thailand (LeFevre 2009). Elephants are also used to haul tourists around on elephant-trekking safaris in both Southeast Asia and increasingly in Africa (more on this in chapter 6).

The use of service animals in the military and as part of law enforcement is, not surprisingly, a long and extensive one. Mary Thurston (2007) documents how in this capacity animals have served as offensive/defensive weapons; as couriers; as test subjects for biological, chemical, and nuclear weapons; as guards; and as transport. She notes that war elephants emerged in India sometime between the eighth and fourth centuries BCE and then diffused to the Persian Empire and on into the Mediterranean. In one of the more famous stories, Hannibal of Carthage, in 215 BCE, traveled over the Alps with a small contingent of war elephants to attack the Roman Empire in what is now Italy. Alexander the Great also acquired war elephants after winning battles in India. Thurston also notes that horses have been used for thousands of years because of their speed and power. In 450 CE Attila the Hun was using cavalry

THE DONKEY: *EQUUS ASINUS* AND *EQUUS HEMIONUS*

The African ass and the Asiatic ass are part of the taxonomic family Equidae of which there are seven species: the Przewalski's horse, the domestic horse, the African ass, the Asiatic ass, and the plains, mountain, and Grevy's zebras. The Przewalski horse is extinct in the wild, and both species of asses and the mountain and Grevy's zebra are endangered. Asses are part of the order Perissodactyla, or odd-toed ungulates, along with tapirs and rhinoceroses. The first horselike animal appeared around fifty-four million years ago during the Eocene period in North America. The first asses and zebras began to appear in Africa about five million years ago.

Asses, like all equids, are herbivores with upper and lower incisors for cropping vegetation and cheek teeth for grinding. Asses are hindgut fermenters, which means that they have only one stomach and no rumen. They can survive on a wide variety of low-quality plant products as the microorganisms in their cecum (a sac between the stomach and large intestine) help break down the tough plant cell walls. Their body weight rests on the middle digit of each hoofed foot making them surefooted on nearly all types of terrain. Asses are normally around six hundred pounds when full grown with the males slightly larger than the females and the Asiatic ass slightly larger than the African. The gestation period is around twelve months, and jennies (females) give birth to one baby. Asses are social animals and live in groups that include a dominant jack (male) and his harem of females. They use a variety of vocalizations to communicate.

The ass was probably first domesticated in Egypt around four thousand years ago and domestic donkeys have been used as draft animals around the world. The word donkey seems to have come into being in the 1700s in England possibly as a slang term for ass, which was coming into its own as slang for the human behind. It seems to be a derivative of dun—a word describing brown. According to the United Nations Domestic Animal Diversity System, currently 189 different breeds of donkeys are recognized for a total of around forty-five million animals. Domestic donkeys range in size from around two hundred to one thousand pounds depending on the breed and are used largely as transport and pack animals. Miniature donkeys originated in Sicily and Sardinia and stand thirty-six inches high or less. They have become increasingly popular animals to have as pets as they are quite docile and easy to care for. Donkeys have quite a cultural history as well. In the Bible a pregnant Mary rode into Bethlehem on a donkey and Jesus rode a donkey into Jerusalem. Politically the donkey has become the national symbol of the US Democratic Party. The origin of this connection goes back to 1828 when opponents of presidential candidate Andrew Jackson labeled him a jackass, and a cartoon of him riding one appeared in 1837, but not until cartoonist Thomas Nast revived the donkey-Democrat connection in an 1870 cartoon did the association stick.

with saddles and foot stirrups, which enabled his troops to have the edge in terms of balance and control. Pigeons have played a role as couriers in many wars throughout history because of their homing instinct. Evidence exists for the use of dogs in the military going back to the ancient Mediterranean cultures where they served as guards, couriers, and attack animals. "Commencement of the 'war to end all wars' in 1914 saw the largest mobilization of animals in history. Three million horses, mules, and oxen; 50,000 dogs; and scores of other creatures were ensnared in this protracted and devastating conflict. World War I would prove fatal for most of these animals because for the first time they were being pitted against mechanized weaponry and lethal chemical agents" (Thurston 2007, 51). Since WWI the use of animals on the battlefield has declined with the rise of more advanced machinery, but we still see large numbers of military dogs being deployed for the same reasons as thousands of years ago. The US and Russian militaries have also harnessed the power of animals such as dolphins, which can detect mines or perform rescue operations in marine environments (Tayman 2012). In terms of law enforcement, dogs and horses continue to be part of police forces around the world. Horses are used for crowd control, and dogs perform search and rescue, sniff for bombs and drugs, chase suspects, and defend officers.

Service and therapy animals are a more modern twist on working animals. Dogs, horses, and even monkeys and pigs are used to help people with disabilities, children/elderly, and the infirm. They are trained to socialize, work gently, remain calm, complete tasks, and stay focused. Therapy animals can be classified into two types. Animals of the first type are owned by volunteers and come to various facilities to allow the patients or residents to interact with the animals. These animals are defined by their duties of animal-assisted activities (AAA). The second type is animal-assisted therapy (AAT) (Center to Study Human Animal Relationships and Environments [CENSHARE] 2011). In AAT the animals are actually tools of a health-care professional in treatment, in order to reach a particular goal. Evidence has been found regarding the positive influence of these animals for humans, providing such benefits as improved motor skills, a reduction of autism symptoms, increased self-esteem, and a reduction in anxiety and loneliness. They can also increase motivation for individuals to get involved in interaction with group activities, and with their caretakers. Service animals are defined as animals that provide direct services to people with disabilities. These helpmates fall into three general classifications: animals that guide the blind, animals that signal the hearing impaired, and animals that perform functions for the disabled. The use of service animals came about in 1929, when The Seeing Eye, an organization that trained dogs for the blind, was established (The Seeing Eye 2011). Signaling dogs came about in the mid-1970s. Service animals are exempted from regular pet laws, and their use is regulated in the United States through the Americans with Disabilities Act (ADA), which became law in 1990 (US Department of Justice [USDOJ] 2002). For example, regulations are in place that allow service animals to be in homes or apartments where pets are otherwise not allowed. While AAA animals may live as pets most of the time and "volunteer" to help people, service animals, or AAT animals, are not pets. They are working animals.

This brief survey of working animal historical geographies allows us to see not only the length of time that working animals have been part of human societies, but also the geographic extent of our ability to do so. In the remaining sections of the chapter we will focus on just one or two specific practices.

ECONOMIC GEOGRAPHIES

For this section we are going to focus on two examples of working animal economies: (1) the economics of animal research and laboratory mice and (2) the horse economy in the United States. An accurate statistic for the number of animals being used in research around the world is very difficult to arrive at. The problems are multifaceted: not all governments require reporting, legal definitions of a research animal differ in different countries (e.g., in the United States rats, mice, birds, and reptiles are not considered animals for the purposes of experimental regulation), and the amount of reporting may or may not be publicly accessible. In an attempt to assess the numbers of animals used, Taylor and colleagues (2008) reviewed animal research publications and triangulated the animals listed in the publications with available government data from 179 countries. They estimated that around 115 million animals are being used for research globally each year. They state the actual number may be much higher, but with a global lack of transparency and adequate reporting, being certain is difficult. They argue that having actual numbers properly broken down is essential to democratic systems, key to advancing science, and helpful in terms of public relations as many people are opposed to, or wary of, using animals in experiments. They conclude that the United States uses the most animals (around 17 million), while Japan uses around 11 million, and the United Kingdom and Germany use just over 1.8 million each. While China does not produce any publicly available statistics, the authors estimated from publications coming out of China that they are using nearly three million animals per year. These animals are being used in a wide range of areas including attainment of basic biological knowledge, human medical research, veterinary research, vaccine and drug development, toxicity testing, and education and training.

Bottini and Hartung (2009) estimate that seventy-three thousand to three hundred thousand people work as direct animal researchers around the world in the areas of medical research (for both humans and animals), basic research (biological studies), and various types of product testing (for food, cosmetics, medicines, chemicals, and products). They also argue that a correlation can be established between gross domestic product (GDP) and animal use—that the larger the GDP the more animals are being used. This conclusion is corroborated by Taylor's analysis as the United States, Japan, China, the United Kingdom, and Germany are the top powerhouses of the global economy. Bottini and Hartung also point out that the biggest producers of animals for experimental purposes—Covance, Charles River Laboratories, MPI Research, Harlan Sprague Dawley, and Huntingdon Life Sciences—average around $5.4 billion in sales annually.

A database search through the International Mouse Strain Resource webpage at Jackson Laboratories returned 3,337 strains of live mice available from around the world for researchers to purchase (many more strains are created within research labs and are often not publicly available) (Jackson Laboratories 2011). Examples of traditional research mice include "stargazer" (due to a defect in the inner ear it is forced to look upward) and "punk-rocker" (a black mouse who constantly bangs its head against its cage). These "normal" lab mice cost around $17 each but a genetically engineered (GE) mouse such as a "GFP transgenic" mouse (a mouse with jellyfish genes that causes it to "glow" under a black light) can cost somewhere between $143 and $193 (Forbes 2001). Profits for these companies are not insignificant. For example, Charles River Laboratories, also known as the "mouse company," had revenues of $465 million in 2001—of which 40 percent came directly from the sale of mouse research models (Aoki 2002). In 2005, the National Institutes of Health (NIH) paid $10 million to purchase 250 strains of GE mice from two companies—Deltagen and Lexicon (Crenson 2006). Chip Murray, the head of intellectual assets for DuPont and supervisor of the Oncomouse patent, commented in 2004 that "we know we have a very important property, and it's in our best interests to get it as widely used as possible" (Jaffe 2004).

The US federal government also supports other mouse researchers and is heavily invested in developing GE mice. For example, recognizing the phenomenal growth in the production of GE mice, agencies are funding

veterinary schools and research institutes to ensure enough animal researchers are being trained to handle this increase (National Research Council
2004). Another area of financial support has come through the sequencing
of the mouse genome, which was completed in 2002. A $130 million effort
funded by a combination of governments (both US and international) and
private sector companies, the sequence was assembled by the International
Mouse Genome Sequencing Consortium (IMGSC) made up of the Whitehead
Institute in Cambridge, Massachusetts, Washington University in St. Louis,
and the Wellcome Trust Sanger Institute and the European Bioinformatics
Institute in England (IMGSC 2002). The sequence shows the order of the
DNA chemical bases A, T, C, and G along the twenty chromosomes of a fe-
male mouse from the "Black 6" strain—the most common laboratory mouse.

Given the fact that mice of all research strains are of such financial and
intellectual value to researchers and businesses, not surprisingly the federal
regulatory systems have contradictory relationships regarding research mice.
Currently, the United States excludes rats, mice, and birds from regulation
under the Animal Welfare Act (AWA)—the main federal law regulating the
treatment of animals, overseen by the US Department of Agriculture (USDA).
This exclusion was made permanent in the 2002 Farm Bill and legally does not
recognize rats, mice, and birds as "animals." Within the NIH, animal welfare
standards do apply to mice under the Public Health Services Act, but only if a
researcher receives a grant. For a private company funding its own research, no
federal laws regulate the treatment of mice. Record keeping for these federal
agencies is also absent. Since mice aren't legally recognized as animals under
the AWA, no records are kept by the USDA as to their numbers in laboratories
or of any pain and distress reports (which are required for recognized species).
The NIH does not publicly track research mice either. Another branch of the
federal government, however, does recognize mice—right now only GE mice.
Literally hundreds of patents have been granted for strains of GE mice by the
USPTO (for a full discussion of patenting living beings, see the political/ethi-
cal section). Conferring a patent gives the patent owner exclusive rights of use
to the organism and the ability to license out the research animal to non–patent
holders—thereby increasing incentive for production.

Housing and experimenting on mice requires a tremendous amount of
laboratory infrastructure. Equipment companies such as Braintree Scientific,
Inc. (of Massachusetts); Lab Products, Inc. (of Delaware); Harvard Apparatus, a Harvard Bioscience's company (of Massachusetts); and PMI Nutrition
International (of Indiana), a subsidiary of Purina Mills, Inc., provide everything from specially irradiated food, to housing units, research equipment,
and equipment for humans (gloves, masks, etc.). Oak Ridge National Laboratory (ORNL) spent $14 million to build a state-of-the-art vivarium, which

can house sixty thousand mice. Only designated researchers and animal-care contractors can enter the Russell Lab. They must take "air showers" to remove debris from their clothing, put on special shoes, and "suit up" before they handle mice. Mouse food, bedding, cages, glassware, surgical equipment, and anything else brought into the facility must be spray-disinfected, fumigated, or sterilized in steam (ORNL 2004). Mapping the commodity chain of lab mice, and especially GE models, reveals a complex human network. Hundreds of millions of dollars and thousands of researchers' careers are invested in maintaining the mouse research laboratory. The development of genetic engineering is actually contributing to an economic boom in the production of new murine research models.

What happens when we turn to the horse economy in the United States? A 2005 report from the lobbying group the American Horse Council (AHC) analyzed the economic contributions of all sectors of human-horse interactions (AHC 2005). These interactions include racing, showing, recreation, ranch work, rodeos, carriage businesses, polo, and police work. They estimated around 9.2 million horses in the United States with 2 million direct horse owners and 4.6 million people involved in the industry as owners, service providers, and employees. These numbers do not include the millions of people who participate as spectators. The council concluded that the horse industry directly contributes $39 billion per year and $102 billion per year when industry suppliers are taken into account. Furthermore the horse industry collectively pays nearly $2 billion in taxes and provides over 700,000 jobs (nearly 460,000 of those full-time). Indeed, they claim that one out of every sixty-three Americans is involved with horses. The sector breakdown is relatively even with racing, showing, and recreation being the primary industry practices.

With just these two examples, laboratory mice and horses, we can see that working animals are a significant contributor to economic activity and that incentive comes from multiple directions—whether from spectators or stockholders—to put animals to work for human benefit.

THE CULTURAL LANDSCAPE

When considering how we encounter working animals in the everyday built environment around us we will concentrate on two areas: zoos and laboratories. Most of us at some point in our lives have visited a zoo, but most of us have probably never been inside an animal research laboratory. In this section we will try to reflect on the similarities and differences between zoos and research labs.

In an article reflecting on early work in animal geography, Michael Watts argues that "zoos inevitably disappoint. The excitement of the wild is replaced by alienation, lethargy, isolation, incarceration and boredom. . . .The inescapable fact is that the zoo recapitulates the relations between humans and animals, between nature and modernity" (2000, 292). Geographers have focused on zoos as spaces that denote the separation of humans from animals as well as on the dimensions that these specific zoo sites reveal about human experiences of the animal other. Most people are used to thinking of zoos as being about entertainment and education; however, animal geography work on zoos asks us to go deeper and ask questions about the role of zoos in the world. What are they telling us about the place of humans and animals?

A key article that exposes us to this animal geography challenge comes from Kay Anderson: "At the zoo, the 18th century ideal of observation via the artifice of distance became popularized in a particular form of visualization and sensory appeal . . . [and] thus the elaborate scene of the metropolitan zoo was constructed, a space in which an illusion of nature was created and re-presented to human audiences in a cultural achievement" (1998, 27). In Anderson's view, the zoo is where "the raw material of nature is crafted into an iconic representation of human capacity for order and control" (31) and exists as "a realm conceived by human imagination for human consumption" (45). Anderson provides a foundation to all animal geography work on zoos because she reveals the ways in which zoos have shaped not only human-animal and human-nature relations but also relations between peoples (cultural groups) and the mediation of knowledge about the world from the prevailing scientific discourse.

So how do zoos go about creating their artificial sites of enclosure? Animal geographers have considered this question in two ways thus far. First of all, Pyrs Gruffudd (2000) shows us how zoos have changed their understandings of proper zoo "habitats" or enclosures. In focusing on the architectural design of early twentieth-century zoos, he highlights how the culture of modernism—one that was about the healthy and efficient nurturing of life—created zoo modernism. "In this sense, the London Zoo was a symbol of the contemporary concern for 'planning' and for a reformed relationship between humanity and nature" (222). Focusing on the designs of Berthold Lubetkin and the gorilla house and penguin pool, Gruffudd shows how Lubetkin adopted a geometric approach to enclosure design because the "architect's task was to understand the animal in its essence and then to coax it through design to display its distinctive characteristics" (225). Thus, he designed a geometric penguin pool in which "penguin-ness" was produced. The enclosure had long ramps for the penguins to waddle on because that was what the humans watching thought of them.

The modern-day zoo differs significantly, in most cases, from the sterile, geometric cages of Lubetkin's time, but the zoo is still evolving. Gail Davies (2000) focuses on the role of new media technologies in mediated experiences at the zoo in order to question whether or not technology helps with the "zoo question"—that is, whether we should keep animals in captivity or not. In using an actor-network approach, she tries "to understand both forms of zoos as complex places, embedded in differing socio-spatial assemblages of people, devices and documents, seeking through different means of representation to make themselves spokespersons for nature" (246). She notes how the history of zoos has been one of change. They have at times been—and continue to be—sites of "collection, colonization, agricultural experimentation, education, and exhibition" (247). As the electronic zoo comes into being with its emphasis on pictures, videos, and digital recordings, Davies argues that perhaps not so much is changing after all. Indeed, in the same way that the traditional zoo inscribes a dominant position for viewers, she believes the electronic zoo does as well because (1) viewers have a more individual rather than collective experience sitting alone in the cinema; (2) the camera has already fixed the animals so they do not have any agency of their own to resist, challenge, surprise, or respond; and (3) animals' previous ability to observe you from their traditional cage has lost all significance. Perhaps electronic zoos are better than incarcerating animals; however, for Davies something is lost when humans no longer interact with real animals—even if the zoo experience does disappoint. While the electronic zoo might be visually stunning, it is devoid of more embodied experiences between species and furthers a mind-set that sees animals as objects of the human gaze—thereby not challenging but instead reinforcing the dualism between subject and object, as well as the separation between humans and animals.

With respect to the cultural landscape of laboratories, two points need to be made. As we can see in figure 4.2, animal laboratories dot the landscape of the United States. Affiliated with universities, biotechnology parks, hospitals, and military facilities, laboratories that experiment on animals literally surround us. Most of us can picture where the farms, zoos, humane societies, and vet clinics are in our hometowns, but how many of us know where the laboratories are? Why is that? The obvious answer is that they are so visible that they disappear into our everyday environments. "Animal research labs are purposefully made invisible for two main reasons—protection of proprietary knowledge and defense against animal rights activists" (Urbanik 2007, 1214). For those concerned about animal research, the wallflower aspect of animal laboratories serves to hide their existence from a public that might otherwise question what goes on inside. Even if we knew where all the labs were, a second point remains to be made about the cultural landscape of

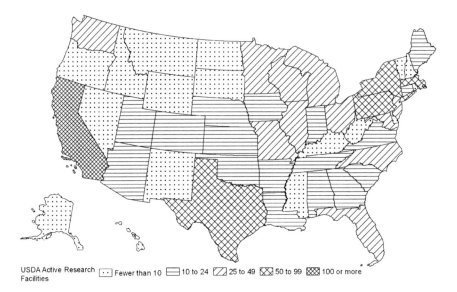

USDA Active Research ⬚ Fewer than 10 ⬚ 10 to 24 ⬚ 25 to 49 ⬚ 50 to 99 ⬚ 100 or more
Facilities

Figure 4.2. Active Animal Research Facilities by State for 2011. *Source:* **Data compiled from the US Department of Agriculture's Animal Welfare database on active licensees and research facilities.**

animal biotechnology specifically. Traditional animal research is conducted in the laboratory or classroom, and stays there. And while this is true for bioengineered animal research models, this is not true for all aspects of animal biotechnology. Animal biotechnology as livestock enhancement, pet production, and endangered animal replacement moves experimental animals outside the laboratory and releases them into our environment to be used to feed us, to provide companionship, or to reshape biodiversity. As a result of animal biotechnology, the scale of the laboratory has expanded dramatically, no longer contained in isolated enclosed spaces. Releasing genetically engineered and altered animals into the environment means that the entire planet has become a laboratory.

So can we compare zoos and animal research labs? In terms of the social construction of lab and zoo animals within our cultural landscape, we can explore four areas. The concept of social construction has to do with a human group's consensus about the meaning of a particular object—both in a physical sense and in a more philosophical manner (Shapiro 2003). This consensus of meaning may or may not be a direct response to the object itself, but, instead, has to do with place. For example, a mouse in a research lab is different from a wild, endangered mouse, which is different still from pet mice or pest mice. With animals, society socially constructs them in certain ways to fit certain ideas about where and how animals should be. The lab mouse is a scientific tool, but the endangered mouse is part of ideas about biodiversity and ecosystem processes; pet mice cannot be treated cruelly but pest mice can be ruthlessly killed at all cost. So what social constructions of lab and zoo animals make them similar? First of all, the construction of these animals as research tools is the same. Research animals are used to solve problems about human health and animal health or to develop and test products. Zoo animals are also research tools because zookeepers use the captive animals to solve problems about animal health, learn about animal behaviors, and practice breeding programs that can be used for conservation purposes.

Second, a loss of individuality occurs in both cases. Research animals are bred specifically for the lab to be exactly like each other in order to maintain scientific consistency for experiments. Zoo animals are often bred specifically for zoos, and while some animals may have names, most of them don't. They are simply Pacific salmon, flamingo, or macaw. Any sense of individuality for animals in these spaces is muted or erased. Christopher Bear uses a case study of Angelica, a giant Pacific octopus that lived at the Deep, an aquarium in Kingston-upon-Hull, in the United Kingdom, to both critique animal geography's tendency to "speak to collectivities such as 'animals,' 'species,' and 'herds,' while speaking less of individual creatures" (2011, 1), and to critique constructions of animals in places such as zoos as interchangeable. In focusing on Angelica, a wild-caught octopus, and her daily life in an aquarium, he teased out more nuanced affective relations between humans and nonhumans. He watched how visitors reacted to her and she reacted to captivity, some days staying still, other days thrashing about her cage, or other times watching her keepers try to enrich her life by designing complex boxes for her to break into to get food. Obviously the keepers attempted to recognize Angelica's individuality, but the most telling part of his research was that during his study Angelica died and was replaced by another octopus. The general public, and even some of the zoo staff, had no idea a different individual was in the tank.

A third constructed similarity has to do with the idea of sacrifice. Supporters of animal research labs argue that the animals are sacrificing themselves for science and for humans so that we can have a better world. They further argue that while animal research might be difficult, the sacrifice of these animals is really the right thing to do and noble. The sacrifice of zoo animals is similar. While zoo animals are not experimented upon for anything other than gaining scientific knowledge about that species, the construction is that sacrificing some number of animals to lifetimes of cages and captivity is acceptable in order to educate humans about the need to keep other members of the same species free and living in the wild. Finally, animals in both cases are constructed as commodities. We have already gotten a sense of the economics of research laboratories, but zoos are also businesses with a need to attract customers, and the commodity is the animals. Whether zoos are selling entertainment, education, a family excursion, or a chance to get close to the wild, they are still selling the animals. We have seen in this section that the cultural landscape of the built environment of zoos and labs creates the physical manifestation of our socially constructed attitudes toward the animals inside these spaces that end up making them more similar than different at the base level.

ETHICAL/POLITICAL GEOGRAPHIES

The political and ethical geographies of working animals have to do with human/animal boundary making and questions about whether using animals for these purposes is ethically permissible. As two examples we will focus on laboratory animals and circus animals. We will begin by returning to Oncomouse, the first patented multicellular living organism. This patent caused a furor among a variety of nongovernmental organizations (NGOs) concerned about the ramifications of having intellectual property ownership of living beings (Rifkin 1998). The scientific community expressed no concern over the patenting of living beings; however, problems have arisen between the owners of Oncomouse and researchers that want to use these models in their work. The high fees charged by patent holders and restrictive licensing agreements have been blamed for impeding scientific progress and intellectual innovation. Patent supporters widely argued that researchers need to maintain ownership of their creations and reap financial rewards because the cost of the research is so high that innovation would not otherwise occur (Jaffe 2004; Raines 1990).

While Oncomouse has been surrounded by legal controversy, the issue of her "mouseness" has never been in question. Humans have kept their human identity intact and decided that these animals count as novel inventions that humans can have control and ownership over; however, the potential development of other forms of transgenic animals and their ability to be patented is raising serious concerns over the clear boundaries between humans and animals and patentable and nonpatentable matter. The "humouse" is an organism imagined by Jeremy Rifkin and Stuart Newman. Both are anti-GE advocates and came up with a way to challenge the legality of patenting living organisms by trying to obtain a patent on a transgenic animal that would be a mix of human and other mammalian embryos. The patent application was actually broken into two—one for a patent on a chimeric organism made of a mixture of human and chimpanzee embryos and the other a mix between humans and mice or other mammals. In 1999, the USPTO rejected the application for the human-chimpanzee patent on the grounds that the organism is human and therefore is not patentable subject matter under the Thirteenth Amendment of the Constitution. The letter of rejection did not state what exactly constituted a human being because the patent office has granted patents on other transgenic animals with human material (Weiss 1999). In August 2004, the patent office rejected the second patent application arguing that the creature "would be too close to human" (Kittredge 2005, 54). Rifkin and Newman believe these patent denials are a victory for those opposed to life patents and hope that they slow the rapid privatization of living organisms. Some researchers, such as Irving Weissman of Stanford University, believe that this patent denial is "a new attempt to block science" (Kittredge 2005, 55). However, the fact that the patent office in both of these patent rejections could not clearly state when a transgenic organism becomes too human to be patentable proves extremely problematic: Does one transgene make the difference (Oncomouse), or is it only when you use human brain cells or human embryonic cells (humouse)?

It is not only anti-GE activists who are uncomfortable with the possibility of creating "human" animals. The National Research Council and the Institute of Medicine of the National Academy of Sciences in the United States recently released a report on using human embryonic cells (National Research Council 2005). The report only presents guidelines and has no legally binding authority; however, it did argue that research into human-chimpanzee chimeras should be halted for the time being. Other chimeric creations were deemed acceptable, but the report expressed the need for adequate oversight committees and close scientific scrutiny of developing research. As an overarching concern, humans have a "strong interest in avoiding any practice that would lead us to doubt the claim that humanness is a necessary (if not sufficient) condition for full moral standing" (Robert and Baylis 2003, 2).

The humouse of Rifkin and Newman was never a real living being, but to obtain a patent the applicant only has to prove the ability to make the patented material. Other researchers, like Weissman of Stanford, have begun developing some of these organisms with problematic identities. Robert and Baylis document four examples of chimeric research:

> Colleagues at Harvard have transplanted human neural stem cells into the forebrain of a developing bonnet monkey in order to assess stem cell function in development; human embryonic stem cells have been inserted into young chick embryos by Benvenisty and colleagues at the Hebrew University of Jerusalem; and most recently it has been reported that human genetic material has been transferred into rabbit eggs, while Weissman and colleagues at Stanford

University and StemCells, Inc. have created a mouse with a significant proportion of human stem cells in its brain. (2003, 1)

If these "human" animals either perceive experiences or behave in human
ways, our obligations toward them will be uncertain. If experimenting upon
human beings is unethical right now, then how would experimenting on a
"humanized" animal be possible? "If we breach the clear (but fragile) moral
demarcation line between human and nonhuman animals, the ramifications
are considerable, not only in terms of sorting out our obligations to these new
beings but also in terms of having to revisit some of our current patterns of
behavior toward certain human and nonhuman animals" (Robert and Baylis
2003, 9). Conceivably ethical recognition of humouse-style organisms could
occur at the expense of organisms like Oncomouse who are not considered
human enough to be a part of the human ethical community.

Gail Davies (2011) examines how technologies are impacting the human-
animal relationship by showing how the making of "monstrous" mice—that
is those with all manner of genetic mutations—reveals the malleability of
not only other species, but, by extension, our own. "If in the 16th and 17th
centuries they [monsters] inhabited the unmapped spaces beyond the known
world, and in the 18th and 19th they arose from the in-between states of
natural history classifications; in the 20th and 21st centuries they seem to
demonstrate a more explicitly political identity, fracturing the humanist assumptions of Enlightenment thought" (1). In arguing that monsters do not
have to look like monsters to have disruptive effects on human society, she
goes inside the laboratories where these mutant and monstrous animals are
being created to understand how, what have always been liminal creatures,
are even more so now that 450 plus genetically modified strains of mice are
marketed. All of this from the Victorian fancy rat and mouse societies that
helped codified breeding practices and eventually led to the inbred strains
of laboratory mice critical to modern science today. Genomic technologies
are pushing this control over breeding and genetic manipulation; however,
these "advances" are fraught with uncertainty. Davies points out how scientists often question the usefulness of so many millions of expendable beings
and that they are concerned that the lack of knowledge of "mouseness" or
"ratness" makes it difficult to distinguish genetic anomalies or draw con-
clusive results.

Work by Beth Greenhough and Emma Roe shows how "there is a conflict
between those who see the experiment from the perspective of animal welfare and those who focus on the research aims and objectives. This conflict
arises from different ways of knowing the experimental subject informed
by different knowledge-practices of a laboratory scientist carrying out an

animal experiment and those of a veterinary expert in animal welfare responsible for ensuring animal welfare in the laboratory" (2011, 54). They document, via interviews with animal researchers, how animal caretakers use critical anthropomorphism to become "experts in interbody communications" by "developing a repertoire of skills that supplement a generalized somatic awareness with species-specific sensitivities developed through time spent with chickens, or monkeys or mice" (55). For the authors, this recognition of the cross-species understanding is a critical component of engaging with a more-than-human world and allowing the animals to be subjects instead of objects.

Even something as seemingly straightforward as animals in higher education has become a political/ethical issue. In a report by Animalearn, the education division of the American Anti-Vivisection Society, which documents the use of cats and dogs in higher education, they argue that (1) no difference can be found between these animals and the dogs and cats that are companion animals, and (2) this use is inhumane and unnecessary given the number of adequate alternatives that exist (Ducceschi and Green 2009). Using a data sample of 175 public colleges and universities, they found that of the 92 universities that responded to their request for information, 52 percent use live or dead dogs and cats, 26 percent use live dogs and cats, and 63 percent of biology departments use dead cats to teach anatomy and physiology. Politically, the major issue is one of student choice: whether they should be forced to participate in this use of animals or be offered alternatives. The first student choice policy at the K–12 level was established in 1985 in Florida and then in California in 1988. According to the American Anti-Vivisection Society, as of 2011 only Florida, California, Pennsylvania, New York, Rhode Island, Illinois, Virginia, Oregon, New Jersey, and Vermont have K–12 student choice laws. In higher education, which is not covered by the student choice laws, individual institutions set their policies. In 1994, Sarah Lawrence College was the first institute of postsecondary learning to adopt a formal student choice policy. However, to date, only twenty-eight colleges and universities have adopted policies. These animals are being obtained from pounds and shelters as well as for-profit breeding companies. Laws are even still in place in Minnesota, Oklahoma, and Utah that require pounds and shelters to surrender unclaimed animals to laboratories.

The circus industry has long been the target of many animal activist groups for their treatment of animals. Groups like People for the Ethical Treatment of Animals (PETA) and the Performing Animal Welfare Society (PAWS) have specific campaigns that target the use of animals in this context. These organizations mainly concern themselves with how the animals are cared for and what they see as abusive treatment and training. Perceived abuses of the animals range from long periods of solitary confinement in small cages, to forced working of pregnant animals, to physical violence. Electric prods are regularly used, and often animals are bound at the feet and forcefully moved around by many humans. This industry, like all others that utilize animals, makes money and creates jobs, but is patronized by spectators who may or may not know what goes on behind the scenes. Animal activists utilize protests when circuses come to towns to try and publicize their view, and they also conduct undercover investigations and release footage of treatment that reinforces their view that the animals are being abused. Activists have been effective at making some legal changes for circuses. More than a dozen municipalities in the United States have banned performances that feature wild animals. Costa Rica, Sweden, Singapore, Finland, India, and Austria ban or restrict wild animal performances, while districts in Australia, Argentina, Brazil, Canada, Colombia, and Greece ban some or all animal acts. These bans have passed because people have decided either that circuses are not the right place for animals or that their treatment in that space is unethical. On the other hand, people have not yet decided that animals are out of place in laboratories.

This chapter has provided an overview of the myriad ways in which humans utilize animals for labor as service, educational, and entertainment animals. Most of these practices have been in existence for hundreds or thousands of years and contribute in various ways to the economies of communities around the world. This prevalence has not meant that these relations exist without question, and animal activists are constantly challenging social constructions that permit animals to exist in places like labs, zoos, circuses, racetracks, and fighting rings.

DISCUSSION QUESTIONS

1. What is it about the urban identity that requires a zoo? Is it less true for nonurban areas? Why or why not?
2. Why aren't there animal research toys for children like there are farm toys, circus toys, rodeo toys, and so on?
3. Do you agree that similarities can be found between zoos and animal research labs?
4. What individual working animals do you know? In what locations? How and why do you know them? In other words, do you have a search and rescue dog or have you just seen people with them?

KEYWORDS/CONCEPTS

animal-assisted activities

animal-assisted therapy

Animal Welfare Act

cultural hearth

humouse

mahout

Oncomouse

service animal

social construction

PRACTICING ANIMAL GEOGRAPHY

1. Visit your local zoo or other entertainment spot (circus/rodeo) and study the messages you get about the animals there. What is being "sold" to you as a consumer/spectator about the animals as a whole and individually?
2. Research and make a map of all the working animals in your area.

RESOURCES

American Anti-Vivisection Society: http://www.aavs.org

American Association for Laboratory Animal Science: http://www.aalas.org

American Association of Zoo Keepers: http://aazk.org

Biotechnology Industry Organization: http://www.bio.org

Eden Consulting Group (police dog training): http://www.policek9.com

Foundation for Biomedical Research: http://www.fbresearch.org

Lolita: Slave to Entertainment (documentary about a wild-caught orca whale that has lived for forty years in Miami): http://slavetoentertainment.com/index2.htm

One Small Step: The Story of the Space Chimps (documentary about the first chimpanzees in space): http://www.spacechimps.com

People for the Ethical Treatment of Animals: http://www.peta.org

Performing Animal Welfare Society: http://www.pawsweb.org

US Department of Agriculture's Animal Welfare Act: http://awic.nal.usda.gov/government-and-professional-resources/federal-laws/animal-welfare-act

World Association of Transport Animal and Welfare Studies: http://www.taws.org

World Association of Zoos and Aquariums: http://www.waza.org/en/site/home

REFERENCES

Advanced Cell Technology. 2001. "ACT Announces Birth of First Cloned Endangered Species." Accessed July 7, 2011. http://www.advancedcell.com/pr_01-12-2001.asp.

AlaskaNet. 1995. "Iditarod History." Accessed April 1, 2011. http://www.alaskanet.com/Tourism/Activities/iditarod/history.html.

American Horse Council. 2005. "National Economic Impact of the U.S. Horse Industry." Accessed February 17, 2011. http://www.horsecouncil.org/national-economic-impact-us-horse-industry.

Anderson, Kay. 1998. "Animals, Science, and Spectacle in the City." In *Animal Geographies*, edited by Jennifer Wolch and Jody Emel, 27–50. New York: Verso.

Animal Planet. 2011. "A History of Horse Racing." Accessed August 2. http://animal.discovery.com/tv/jockeys/horse-racing/history.

Aoki, Naomi. 2002. "The Mouse Company Roars." *Boston Globe*, May 21, Business Section, E6.

Applebome, Peter. 1989. "Wrangling over Where Rodeo Began." *New York Times*, June 18. Accessed August 17, 2011. http://www.nytimes.com/1989/06/18/travel/wrangling-over-where-rodeo-began.html?pagewanted=all&src=pm.

Bear, Christopher. 2011. "Being Angelica? Exploring Individual Animal Geographies." *Area* 43 (3): 297–304.

Bottini, Annamaria A., and Thomas Hartung. 2009. "Food for Thought . . . on the Economics of Animal Testing." *ALTEX* 26 (1): 3–16.

Bumiller, Elisabeth. 2011. "The Dogs of War: Beloved Comrades in Afghanistan." *New York Times*, May 2, A12. Accessed May 2, 2011. http://www.nytimes.com/2011/05/12/world/middleeast/12dog.html?_r=1&ref=unitedstatesspecialoperationscommand.

Center to Study Human Animal Relationships and Environments. 2011. "Animal Assisted Therapies." Accessed October 12. http://censhare.umn.edu/AAT.html.

Chambers, Paul. 2009. *Jumbo: This Being the True Story of the Greatest Elephant in the World*. Hanover, NH: Steerforth Publishers.

Clark's Trading Post. 2011. "Welcome to Clark's Trading Post." Accessed July 17. http://www.clarkstradingpost.com.

Crenson, Matt. 2006. "Mice Are Key Tool in Quest for New Drugs." Accessed June 3, 2011. http://www.abcnews.go.com/US/print?id=1688344.

Davies, Gail. 2000. "Virtual Animals in Electronic Zoos: The Changing Geographies of Animal Capture and Display." In *Animal Spaces, Beastly Places: New Geographies of Human-Animal Relations*, edited by Chris Philo and Chris Wilbert, 243–267. New York: Routledge.

———. 2011, April 23. "Writing Biology with Mutant Mice: The Monstrous Potential of Post Genomic Life." *Geoforum.* Published electronically. doi:10.1016/j.geoforum.2011.03.004.

Ducceschi, Laura, and Nicole Green. 2009. *Dying to Learn: Exposing the Supply and Use of Dogs and Cats in Higher Education*. Jenkintown, PA: American Anti-Vivisection Society.

Emel, Jody, and Julie Urbanik. 2005. "The New Species of Capitalism: An Ecofeminist Comment on Animal Biotechnology." In *A Companion to Feminist Geography*, edited by Lise Nelson and Joni Seager, 445–457. Oxford: Blackwell.

Forbes. 2001. "How Much Is the Glowing Monkey in the Window?" Accessed May 23, 2005. http://www.forbes.com/2001/02/02/0202transgenic_print.html.

Gibson, Hanna. 2005. "Dog Fighting Detailed Discussion." Accessed August 20, 2011. http://www.animallaw.info/articles/ddusdogfighting.htm.

Greenhough, Beth, and Emma Roe. 2011. "Ethics, Space, and Somatic Sensibilities: Comparing Relationships between Scientific Researchers and Their Human and Animal Experimental Subjects." *Environment and Planning D: Society and Space* 29:47–66.

Gruffudd, Pyrs. 2000. "Biological Cultivation: Lubetkin's Modernism at London Zoo in the 1930s." In *Animal Spaces, Beastly Places: New Geographies of Human-Animal Relations*, edited by Chris Philo and Chris Wilbert, 222–242. New York: Routledge.

Hancocks, David. 2001. *A Different Nature: The Paradoxical World of Zoos and Their Uncertain Future*. Berkeley: University of California Press.

International Circus Hall of Fame. 2011. "History of the Circus." Accessed July 7. http://circushof.com/circus_history.html.

International Mouse Genome Sequencing Consortium. 2002. "The Mouse Genome and the Measure of Man: Press Release." Accessed November 1, 2010. http://www.broad.mit.edu/media/press/pr_02_mousegenome.html.

Jackson Laboratories. 2011. "International Mouse Strain Resource (IMSR) Database." Accessed August 15. http://www.informatics.jax.org/imsr.

Jaffe, Sam. 2004. "Ongoing Battle over Transgenic Mice." *The Scientist* 18 (14): 46.

Jayasekera, Palitha, and Shelton Atapattu. 1995. *Elephants in Logging Operations in Sri Lanka*. Rome: United Nations Food and Agriculture Organization.

Kittredge, Clare. 2005. "A Question of Chimeras." *The Scientist* 19 (7): 54–55.

Kolata, Gina. 1998. *Clone: The Road to Dolly and the Path Ahead*. New York: William Morrow.

LeFevre, John. 2009. "Thailand's First Retirement Home for Elephants Opens in Lampang." Accessed August 2, 2011. http://www.thailand-travelonline.com/thailand-destinations/northern-thailand-information/thailands-first-retirement-home-for-elephants-opens-in-lampang/1578.

Loew, Franklin M., and Bennett J. Cohen. 2002. "Laboratory Animal Medicine: Historical Perspectives." In *Laboratory Animal Medicine*, 2nd ed., edited by James G. Fox, Lynn C. Anderson, Franklin M. Loew, and Fred W. Quimby, 1–17. New York: Academic Press.

National Research Council. 2004. *National Need and Priorities for Veterinarians in Biomedical Research*. Washington, DC: National Academy of Science.

———. 2005. *Guidelines for Human Embryonic Stem Cell Research*. Washington, DC: National Academy of Science.

Nichols, Ashton. 1999. "Romantic Rhinos and Victorian Vipers: The Zoo as Nineteenth-Century Spectacle." Accessed April 12, 2011. http://users.dickinson.edu/~nicholsa/Romnat/zoos.htm.

Norton, Bryan G., Michael Hutchins, Elizabeth F. Stevens, and Terry L. Maple, eds. 1995. *Ethics on the Ark: Zoos, Animal Welfare, and Wildlife Conservation*. Washington, DC: Smithsonian Institution Press.

Oak Ridge National Laboratory. 2004. "A Clean Mouse Research Lab." *Oak Ridge National Laboratory Review* 37 (3): 10.

PBS. 2011. "A History of the Circus." Accessed July 14. http://www.pbs.org/opb/circus/in-the-ring/history-circus.

Rader, Karen. 2004. *Making Mice: Standardizing Animals for American Biomedical Research, 1900–1955*. Princeton, NJ: Princeton University Press.

Raines, Lisa. 1990. "Public Policy Aspects of Patenting Transgenic Animals." *Theriogenology* 33 (1): 129–149.

Rifkin, Jeremy. 1998. *The Biotech Century: Harnessing the Gene and Remaking the World*. New York: Jeremy P. Tarcher/Putnam.

Robert, Jason, and Francoise Baylis. 2003. "Crossing Species Boundaries." *American Journal of Bioethics* 3 (3): 1–13.

Rose, Mark. 2010. "World's First Zoo—Hierakonpolis, Egypt." *Archaeology* 63 (1): 25.

The Seeing Eye. 2011. "Historical Timeline." Accessed October 17. http://www.seeingeye.org/aboutUs/default.aspx?M_ID=472#1970%27s.

Shapiro, Kenneth J. 2003. "A Rodent for Your Thoughts: The Social Construction of Animal Models." In *Animals in Human Histories: The Mirror of Nature and Culture*, edited by M. Henniger-Voss, 439–469. New York: University of Rochester.

Space Today. 2011. "Monkeys and Other Animals in Space." Accessed June 14. http://www.spacetoday.org/Astronauts/Animals/Dogs.html.

Stagecoach Inn Museum. 2011. "Jungleland of Thousand Oaks." Accessed July 14. http://www.stagecoachmuseum.org/jungleland_exhibit/jungleland_exhibit.htm.

Taylor, Katy, Nicky Gordon, Gill Langley, and Wendy Higgins. 2008. "Estimates for Worldwide Laboratory Animal Use in 2005." *Alternatives to Laboratory Animals: ATLA* 36:327–342.

Tayman, Regina. 2012. "Navy Dolphins to Keep the Strait of Hormuz Open." Accessed February 9. http://www.milpages.com/blog/1601815.

Thurston, Mary. 2007. "Animals in War." In *Encyclopedia of Human-Animal Relationships: A Global Exploration of Our Connections with Animals*, edited by Mark Bekoff, 51–56. Westport, CT: Greenwood.

Urbanik, Julie. 2007. "Locating the Transgenic Landscape: Animal Biotechnology and Politics of Place in Massachusetts." *Geoforum* 38 (6): 1205–1218.

US Department of Justice. 2002. "ADA Business BRIEF: Service Animals." Accessed August 8, 2011. http://www.ada.gov/svcanimb.htm.

US Patent and Trademark Office. 1998. "Transgenic Non-human Mammals." Accessed July 7, 2011. http://patft.uspto.gov/netacgi/nph-Parser?Sect2=PTO1&Sect2=HITOFF&p=1&u=/netahtml/PTO/search-bool.html&r=1&f=G&l=50&d=PALL&RefSrch=yes&Query=PN/4736866.

Walker, Sally R. 2001. "Acquisition." In *Encyclopedia of the World's Zoos*, edited by Eatharine E. Bell, 1–3. Chicago: Fitzroy Dearborn.

Watts, Michael. 2000. "Afterword: Enclosure." In *Animal Spaces, Beastly Places: New Geographies of Human-Animal Relations*, edited by Chris Philo and Chris Wilbert, 292–304. New York: Routledge.

Weiss, Rick. 1999. "U.S. Ruling Aids Opponent of Patents for Life Forms." *Washington Post*. Accessed August 1, 2011. http://pqasb.pqarchiver.com/washingtonpost/access/42476319.html?FMT=ABS&FMTS=ABS:FT&date=Jun+17%2C+1999&author=Rick+Weiss&pub=The+Washington+Post&edition=&startpage=A.02&desc=U.S.+Ruling+Aids+Opponent+Of+Patents+for+Life+Forms

WinningPonies. 2011. "Horse Racing History." Accessed February 1. http://www.winningponies.com/horse-racing-history.html.

Chapter Five

Down on the Farm: Geographies of Animal Parts

Have you ever eaten a lion-meat taco? Would you? Most people would not think twice about eating a cow, chicken, or fish taco, but a lion taco? In early 2011, Boca Tacos y Tequila of Tucson, Arizona, advertised that it was going to make and sell lion-meat tacos. While lions are a threatened species in the wild, in the United States raising and selling lions for meat is perfectly legal under the Federal Department of Agriculture regulations because the animals are not classified as endangered, and they are raised in captivity (Vinyard 2011). The publicity stunt caused a huge wave of public resistance, and the restaurant backed out of the lion-meat taco business and went back to cows, chickens, and fish.

The World Cup soccer games in Seoul, Korea, were a lightning rod for animal activists because of the small percentage of Koreans who eat dogs (Chaudhary 2002). Since the Western world sees dogs more as pets than as food, organizers were concerned both sides would not be able to bridge the cultural divide. Animal activists said they protested the ways in which the dogs are treated: they are often beaten to death or die slowly because the Koreans say it makes the meat taste better. Koreans claim that eating dogs is no different than eating a lamb or a pig, and that the disagreement was all about cultural practices.

A law going into effect in 2012 in California will make it illegal to sell foie gras—a paté made of the enlarged livers of force-fed geese. Enough people in the state found this practice to be objectionable to pass a ban against the wishes of those who enjoy foie gras as a delicacy and those who believe they should be able to choose what they eat (Nagourney 2011).

Legislators in the state of Iowa recently attempted, but failed, to pass a bill that would make it a crime to produce, distribute, or possess photos or videos taken without permission from agricultural facilities (Sulzberger 2011).

Those in the agricultural industry are concerned that this favored tactic of animal advocacy groups harms their business images and becomes costly if a company loses customers and incurs fines for violations of food safety and animal cruelty laws when such events are isolated to specific employees and not standard practice. Proponents of the right to know believe this exposure is essential for monitoring the ways our food is being produced both for animal welfare reasons and for human health reasons. Indeed, in many instances these exposés have forced recalls of eggs or meat because authorities were alerted to unsanitary practices. The history of this type of undercover work goes back over one hundred years to Upton Sinclair. In his 1906 novel *The Jungle*, he depicted the lives and conditions of Chicago's meatpacking district revealing the horrific conditions of the workers and animals, and ultimately the negative consequences for the human consumers (Sinclair 1906). The uproar caused by this book played an instrumental role in the development of the Federal Meat Inspection Act and the Pure Food and Drug Act in 1906, which, in turn, led to the creation of the federal Food and Drug Administration. Modern-day images of animals being kicked, beaten, and stomped on or animals disease-ridden or dead in their cages continue to have an impact whenever they are aired. The concerns remain the same—the treatment of food animals and workers and the impact on human health. Some people are concerned with "food karma"—they don't like the idea of eating abused animals. Others are bothered by the idea that they could die from eating animal products contaminated with salmonella or E. coli, or that they are ingesting meat heavily laden with antibiotics. Still others are disturbed by the poor pay, long hours, and increasing reliance on illegal immigrant labor to do the country's "dirty" work of killing animals.

The issue of what types of animals we eat and how we eat them has been a culturally significant aspect of human-animal relationships for millennia. The first records of using domesticated animals for food go back at least ten thousand years with different cultural hearths for different species. Domestication, the process of selective breeding, originated for cattle in southwest and south Asia, for chickens in southeast Asia, and for pigs in southwest Asia. Over these millennia we have developed four main methods of producing food animals. Nomadic pastoralism involves people moving across the landscape with their animals (normally cattle, goats, and sheep) looking for the best grazing and water sources. Today, nomadic pastoralists can be found in Africa, the Middle East, and central Asia. The other three methods of raising livestock are sedentary—that is both the animals and people remain in one place. Subsistence animal farming involves raising food animals only for oneself and family. For example, keeping backyard chickens is becoming more popular in urban areas today. In small-scale market farming, businesses either sell surplus products or raise small numbers of animals for the market (possibly up to several hundred). In the newest method of livestock farming, the industrial method, economies of scale allow more animals to be raised in more intensive conditions. For example, usually around 250,000 laying hens are housed in each industrial egg farm building. Industrial animal farmers call these farms concentrated animal feeding operations (CAFOs). The vast majority of the fifty billion land animals killed on the planet each year come from CAFOs (Steinfeld et al. 2006). Industrial animal agriculture is not limited to land animals, and the recent arrival of modern, industrial aquaculture confines and raises marine organisms like fish, shrimp, and oysters.

Food animals are not the only farmed animals in the world. Tigers and bears are farmed mainly in China and Southeast Asia for their body parts—some for food, some for what is considered traditional medicine. In addition, species such as lynx, chinchilla, fox, and mink are farmed for their fur in Asia, Europe, and North America. The animals that humans choose to farm make for fascinating geographies. From the politics of states instituting a ban on images of CAFOs to the literal landscape of the location of the food animals to the slippery ethical terrain and different cultural practices, this umbrella category of farmed animals is one of the key facets of the human-animal relationship. In this chapter we will explore how geographers have looked at this topic as well as gain an understanding of what farming animals in today's world actually means.

HISTORICAL GEOGRAPHIES

Interest in the history of livestock domestication has been long standing in geography. While the complete details of this research are too extensive to cover here, we will highlight some major contributions of the three waves of animal geography. Christine Rodrigue (1992) points out how early modern geographers—loosely correlating with the first wave of animal geography—thought that domestication was a by-product of the need for ritual sacrifice of animals to appease religious beliefs after plant domestication allowed people to live more sedentary lives and spend time developing their cultures. While her examination of the archeological evidence at Neolithic sites in the present-day Middle East yielded no evidence that ritual practices were the dominant force, first-wave animal geographers sill laid the groundwork for understanding spatial variations of livestock at the time.

Frederick Simoons (1974) and James Baldwin (1987) compiled work on what we would now consider second-wave animal geography. They focused on understanding the time and place of livestock domestication, dispersal of

breeds, animals as agents in vegetation change, dairying and lactose intoler-
ance, pastoralism, and livestock diseases—in essence the cultural ecology of
livestock animals. Debates at this time on the origins of domestication had to
do with those who believed humans began domesticating animals for cultural
reasons like economics and religion and those who looked to environmental
pressures such as desertification or glacial advances forcing humans to control their food supply. A few geographers, such as Frederick Zeuner (1963),
were also arguing that domestication may have come about because of more
fundamental social relations between humans and certain wild animals—perhaps affection and pets. The goal of research at the time was to "determine
the effects that using or having to use the animals has upon cultural patterns
and social organization" (Simoons 1974, 568). As an example Simoons turns
to the rise of the "sacred cow" in India and how this notion gained prominence gradually because of the power of the high-caste Brahmins reacting
against the subsequent Muslim invasion of India—this solidified the "sacred
cow" and avoiding beef for the purpose of maintaining cultural differences.
He notes that this idea of protecting cattle differs completely from the West
where we have no sense of cattle as sacred and merely use them as a product.
Paul Robbins (1998) comes to a similar conclusion when he examines the
complex history of vegetarianism in India. While the Hindu religion has a history of practicing ahimsa—or nonviolence—that results in vegetarian practices to avoid committing violence against other life-forms, further research
reveals that India actually has a similarly long history of meat eating by different groups for different reasons. In fact, "the Vedic [Hindu sacred texts]
prescriptions of nonviolent behavior and ritual purity begin to seem less like
a universal cosmology and more like an ideological system of status legiti-
mation" (231). For Robbins the use of vegetarianism as a form of practicing
ahimsa ends up applying to select groups (usually those with less sociopolitical power) while meat eating by the more powerful becomes legitimized.

Kay Anderson (1997) provides the foundation for third-wave animal geography by portraying animal domestication in terms of larger social processes.
She maps out how domestication "frames relations that extend beyond animals to include other human groups encountered as people inhabit and move
about the world" (464). Animals such as goats, pigs, sheep, and cows were
"brought into socially embodied form" and became "hybrids of 'culture' and
'nature'" (465). For earlier geographers, domestication came about because
of advances in culture that made humans exceptional and able to exploit the
natural world like no other animal group. For Anderson, these views make
sense given the history of Western thought toward the natural world, and
she argues that for all of recorded history the ability of humans to modify
other life-forms has been seen as a marker for "humanness." For the Greeks,

domestication was how they demonstrated their rise from animal nature, and under Christianity the separation and elevation of humans above animals and the rest of the environment exemplified humans as God's chosen ones and also allowed humans to rise above their animal nature. This view continued all the way through the Scientific Revolution, which brought rationality to the fore of unique human characteristics (well, some humans), and this rationality was also used to justify human dominance over other species. Anderson challenges these narratives and sees "domestication as a political activity historically interconnected with ideas of human uniqueness and dominion, savagery and civilization that became woven into the structuring of not only human-animal relations but also other social arrangements" (470).

This notion of human exceptionalism, or the ability to "transcend the beast within" (Anderson 1997, 473), and justifiable human dominance over other species translated in the eighteenth century to *Western* exceptionalism and justifiable dominance of *Western* peoples over the lands, animals, and peoples of the rest of the world. Anderson shows the mind-set of European powers while colonizing—a goal of bringing the uncivilized people to civilization to "improve" them and "fix" them—was applied to breeding animals and plants. The fundamental, and radical, goal of Anderson's work is to point out how domestication of animals, far from being some transhistorical evolutionary or God(s)-given right, is, in reality, "a political activity embedded within concrete human practices" (479). The political activity she refers to is the methods or justifications for bringing animals closer to humans and further from the wild yet keeping them constructed as less than human and therefore subordinate—a variant of which was applied to other peoples as well.

Two case studies support Anderson's conclusions about people, power, and domestication. Carl Griffin (2012) documents the intimate and often violent interactions between farmworkers and farm animals at the turn of the nineteenth century in England. In this time of shifting livestock practices, animal production was increasing and money was to be made, but not for the vast majority of farmworkers. This time period also saw the beginning of animal protection in England with the founding of the Royal Society for the Prevention of Cruelty to Animals (RSPCA) and the first animal welfare legislation. In many cases the only power farmworkers felt they had was over the animals, and this feeling of powerlessness against other humans translated into violence toward animals. Violent bestiality and maiming were the extreme results of rural workers being "reduced to bodies, as instruments of capital, as *things* to be regulated. They were 'bare life.' Agrarian capitalism leveled working men and women to the same status as animals" (Griffin 2012, 13). Griffin emphasizes that not all farmworkers participated in violence, and indeed many had close, caring relationships with the animals they owned, but his case study shows how agrarian capitalists were "domesticating" farm laborers by controlling their pay, work habits, and livelihoods in addition to controlling the animals themselves. The manner in which one human group (agrarian capitalists) maintained power over another human group (laborers) even as both had power over the livestock exemplifies Anderson's claim that domestication is about more than just human-animal relations.

Diana Davis (2008) writes about the role of veterinary medicine in aiding the French and British as they colonized North Africa and India in the 1800s and early 1900s. As we already noted in chapter 4, animals were heavily used as military apparatus all the way through World War I. Since the processes of colonization were carried out with military force, healthy animals were a necessity. While the colonizing forces brought their own animals with them, they also adapted to local conditions—the British learning to utilize elephants and the French, camels. Veterinarians were needed to care for animals, study and treat diseases, and, in the case of the French, help with livestock management practices. Interestingly, while farmworkers in England were being exploited, the colonists of India were controlling both animals and Indians in exploitative ways. The British banned nomadic pastoralism and prevented many Indians from practicing their traditional forms of livestock rearing. The French, on the other hand, made use of Algerians as laborers and developed an extensive export industry based on grazing animals like goats and sheep. Shawn Van Ausdal's (2009) work on cattle ranching in Colombia also demonstrates colonialism and the link to environmental and labor changes. He notes that while many people think that deforestation in South America for ranching is a modern-day phenomenon, it actually has much deeper roots and goes back to the 1850s. In these cases we can begin to see how the domination of livestock animals was used to control (and challenge that control of) the environment, animals, *and* people.

A glimpse into more modern historical geographies of domesticated animals comes from work by Chris Philo (1998) on livestock in the city. While Anderson (1997) points out how the Industrial Revolution and growing urbanization brought the domesticated wild into the city via breeding pigeons, fancy mice, and the rise of Victorian pet culture, Philo shows how simultaneously livestock animals were being moved away from the urban areas because they were no longer seen as being in the right place. In Victorian London, concerns about social protocol and modesty were of paramount importance, and herds of livestock animals moving through the streets to be slaughtered became an affront to "civilization." People were concerned about safety (loose animals could cause harm), morality (children and women being exposed to animals having sex or relieving themselves in public), and the stress brought on by witnessing the cruelty of animal drivers and the slaughter process. For Philo this movement marks the beginning of the separation of Western humans from the animals they consumed. People no longer wanted to be exposed to these processes and rather than reconsider them, they simply banished the animals (and thereby their treatment) to spaces where they would be invisible. In the United States, this time period also coincided with western expansion and the removal of native peoples and bison in order to provide rangeland for Americans (Emel 1998). Also in the United States during the late 1800s, the livestock system was being restructured to feed urban areas as the animals were being moved out. Gary Fields (2003) documents how G. F. Swift was able to harness the new technology of refrigeration and adapt it to rail cars so he could ship parts of animals around the country rather than whole live ones. This advance helped reduce costs and allowed for more efficient methods of feeding the cities without people having to know about the lives of the animals.

Even as the animals were being moved out of urban areas, control over animals via breeding was really coming into prominence. Walton (1984) documents how the "improved" shorthorn cow came from eighteenth-century provincial obscurity to a dominant position in the nineteenth century in the United Kingdom. "As a dual purpose animal, suitable for both fattening and milk production, the shorthorn was both without rival among other major breeds and well attuned to the market conditions of the 19th century" (23). Out of this focus on the shorthorn came the world's first herdbook of registered pedigrees. Walton discusses how the herdbook served to highlight class differentials as the ability to breed shorthorn improvements and register them was limited to the aristocracy, but some of the changes did move into the breed as a whole no matter who owned the shorthorns.

The ability to control animal bodies has reached new heights with present-day industrial practices, which have changed dramatically from the earliest days of domestication. Throughout history, access to water and food, the ability to protect and control a herd of animals, and traditional selective breeding practices limited the global livestock population. In the post-WWII period, however, the Western world changed the way animals were raised. The development of industrial animal agriculture was possible because of advances in technology that allowed more animals to be confined in smaller areas, medicines that kept them from getting sick, and breeding that allowed for production on a larger scale (Watts 2000). What Michael Watts refers to as "modern-day acts of enclosure" not only spatially confine animals but also have solidified corporate control over the livestock industry thereby changing the structure of livestock rearing from one of local control to a more fully integrated and homogenized system.

Livestock farming is not the only type of farming that goes on in the world today. Farms that raise animals for their parts also play a significant role. Bear bile has been used in Chinese medicine for thousands of years. Bears are the only mammal to produce ursodeoxycholic acid (UDCA), and although today it can be produced synthetically, demand remains for the natural ingredient. Surplus amounts of bear bile may also be used in shampoos, tonics, and other commercial products. Bear farming for bear bile began in the 1980s and an estimated twelve thousand Asiatic black bears are used for this purpose in China, Vietnam, and Korea (Animals Asia 2011; Hobson, 2007). Sun bears may also be used, but almost always farms cultivate the black bears, or "moon bears." Bear bile farms keep bears in wire cages where they can barely turn around. They have permanent catheters in their gall bladders and the bile is harvested when needed. Bears can be kept alive for years in these conditions. According to the Animals Asia organization, China is home to the highest number of farmed bears (seven thousand to ten thousand) today. Consumers are not just after their gall bladders. Bear teeth, claws, and even paws—for bear paw soup—can fetch high prices in the Asian markets.

According to the World Wildlife Fund (WWF), at least five thousand tigers have been bred in China's tiger farms, more than all the members of the six subspecies combined still in the wild today. Tiger farming continues even though China's trade of tiger parts became illegal in 1993 when the nation issued a ban and removed tiger bone from traditional Chinese medicine (TCM) ingredients (WWF 2011). In Vietnam, Laos, and Thailand tiger farms sometimes openly operate under the guise of conservation. Every part of the tiger is desirable for one reason or another, and some groups will even make tiger wine out of tiger bones. Proponents argue that having tigers on farms protects their miniscule numbers in the wild while opponents argue that consuming tiger at all is unnecessary and contributes to declining wild populations by encouraging poaching. Finally, millions of animals are kept on fur farms predominantly in North America, Europe, and Russia. In this case the animals are kept alive only long enough to obtain their pelts for use in the fashion industry. Most animals die from being electrocuted so the coat is not ruined, and ground-up animals are normally fed back to the living ones.

While historically the use of animals for clothing was done for survival with some attention to fashion and rank or wealth, in today's world furs are largely seen as a display of wealth rather than need. According to the organization In Defense of Animals (IDA), "to make one fur coat you must kill at least fifty-five wild mink, thirty-five ranched mink, forty sables, eleven lynx, eighteen red foxes, eleven silver foxes, one hundred chinchillas, thirty rabbits, nine beavers, thirty muskrats, fifteen bobcats, twenty-five skunks, fourteen otters, one hundred twenty-five ermines, thirty possums, one hundred squirrels, or twenty-seven raccoons" (IDA 2011). One of the main arguments across the board for fur farms follows those for tiger and bear farming—that if we are purposely raising and harvesting these animals we are ensuring their sustainability in the wild.

As we have seen, the past ten thousand years have been busy for humans as they have worked around the planet to harness animal life for human existence. In the next section, we will examine the impact of this use of animals on the economy.

ECONOMIC GEOGRAPHIES

The economics of livestock systems is, like the history, an enormously complex issue and one that cannot be explored fully here. However, we will highlight three ways the livestock industry can be examined from a geographical perspective: the economics of consumption, the animals themselves, and the impact of the industry on specific places. With the first theme, the economics of consumption, we can gain a sense of the scale of the global livestock industry by looking at the major consumers and producers. Figures 5.1 through 5.4 compile data from the United Nations Food and Agriculture Organization (FAO), whose purpose is to gather international data on food systems. What

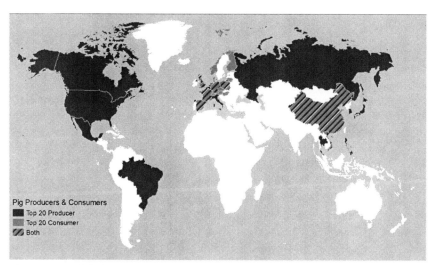

Figure 5.1. Global Pig Production (in metric tons) and Consumption (in kilocalories/person/day). *Source:* Data compiled for the years 2003–2009 from the United Nations Food and Agriculture Statistical Yearbook for 2010.

we can see from these maps is the spatial density of consumption patterns for different animals. In terms of production this data translates to over fifty billion land animals per year with around ten billion animals being consumed in the United States alone (mainly in the form of chicken). The FAO estimates these numbers will double by 2050 due to rising economic standards around the world. Interestingly, as people's incomes rise, one of the first things they begin doing is consuming more meat because they can afford to.

For citizens in industrialized countries who have been consuming large amounts of meat products, the economic issue is not one of access but of health and consumer identities. While many analyses of food commodity chains construct the process as one that is unidirectional in nature from the space of the farm to the kitchen table, Stassart and Whatmore ask us to consider that "the metabolic impressions that the flesh of others imparts to our own are an enduring axiom of social relations with the nonhuman world and the porosity of the imagined borders which mark 'us' off from 'it'" (2003, 449). They are basically saying that we are more connected to our animal-based food products than we realize, and an intercorporeality—a shared bodily experience—of food consumption mixes together consumer and producer knowledge practices. Through a case study of Coprosain they show how one company has shifted its practices—the methods of production and the advertising—to "render the connections between the treatment of living

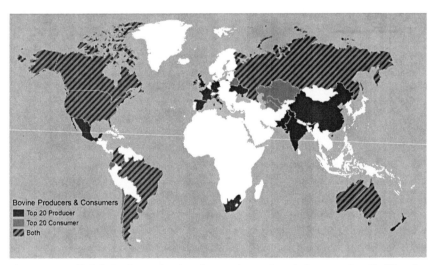

Figure 5.2. Global Cattle Production (in metric tons) and Consumption (in kilocalories/person/day). *Source:* Data compiled for the years 2003–2009 from the United Nations Food and Agriculture Statistical Yearbook for 2010.

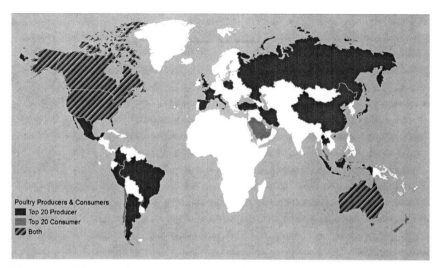

Figure 5.3. Global Poultry Production (in metric tons) and Consumption (kilocalories/person/day). *Source:* Data compiled for the years 2003–2009 from the United Nations Food and Agriculture Statistical Yearbook for 2010.

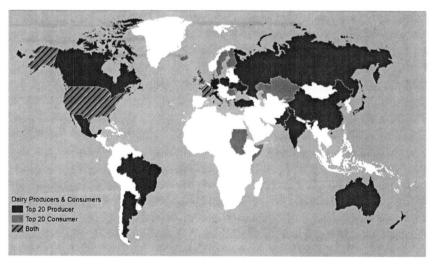

Figure 5.4. Global Dairy Production (in metric tons) and Consumption (in kilocalories/person/day). *Source:* Data compiled for the years 2003–2009 from the United Nations Food and Agriculture Statistical Yearbook for 2010.

animals and the quality of their meat transparent" (451). In making the pig
and its meat effectively become the "spokes-product" for their corporation,
Coprosain reclaimed the animals from industrial invisibility and created a
provisional coming together of animal, producer, and consumer that enables
the company to market itself as a trustworthy provider of quality meat products—ones that are safe both for animals and humans.

The intercorporeality of food and concerns about health can also be found
with aquaculture. Becky Mansfield (2011) demonstrates how aquaculture can
create more problems than it solves by producing "multiple natures." The rise
of aquaculture as an industry is a response to the decline of global wild fish
populations. "This 'blue revolution' is promoted in the name of progress,
development, and increased food production to feed a growing population"
(419). As Mansfield points out, consumers want more and cheaper fish so in
order for the industry to supply it, producers literally have to make a different
fish—one that comes with questionable human health issues. The way they
change the fish is not only by keeping them in cages, but by introducing many
different types of chemicals. These chemicals help them stay healthy, keep
the water clean, increase growth, and produce certain flesh colors (farmed
salmon must be dyed pink because the natural pinkish hue comes from their
diet in the wild). What Mansfield points out, then, is a conundrum for all
industrial animal production—in the drive to meet consumer demand are we
consuming the animals we think we are?

This question leads us directly to our second theme: the construction of the
animals themselves. Changes in technology and economic globalization in
the post-WWII period have led to a restructuring of the animals themselves.
In the case of pigs, restructuring has happened in three ways according to
Frances Ufkes (1998). Her research into the industrialization of the US pork
industry highlights how since the mid-1970s the focus of the industry has
been on greater scale of production and the development of "boutique" pork
products such as lean bacon. This focus results in a more consolidated industry that is vertically integrated with only a few companies (like Smithfield
Farms) dominating the sector. Second, these changes have literally changed
the bodies of the pigs themselves. Increased technological intervention into
genetics, feed, and confinement has meant that the animals themselves are
being homogenized and altered to fit the needs of consumers wanting cheap
meat *and* of producers needing standardized animals to control the feed and
keep the slaughter process efficient. Finally, livestock raising and slaughtering was "a high-wage sector in the early postwar era, [but] food processing
turned into one dependent upon low-wage, rural, female and immigrant
workers" (243). But this shift in pig production has also changed the overall
emphasis of the US market in terms of globalization. Pig consumption, even

though the industry launched the "other white meat" campaign, has not continued to grow as producers had hoped. This lack of growth has led to the globalization of the US pork industry so that products are marketed around the world.

Work by Gibbs and colleagues argue that "the livestock industry is moving towards domination by corporate interests as it becomes part of a wider international bioeconomy" (2009, 1041). We can understand the bioeconomy as that which uses biological information to produce products for the market (e.g., biotechnology). One way the livestock industry is doing this is by supplanting more traditional, hands-on evaluation of animal bodies with scientific evaluation by specialists armed with estimated breeding values (EBVs) software that quantifies an animal's genetic value based on the processing of genetic information about their bodies. They can use the EBVs to market improved breeding and performance of individual animals, sell genetic markers from individual DNA analysis, and use the genetic material to further genetic modification or cloning goals. In essence, the scale of control over the animal bodies is becoming more intense at the same time that genetic practices are becoming privatized and based more on the creation, storage, manipulation, and deployment of data in corporate and state laboratories rather than on the farm with farmers.

Finally, livestock production systems have had an impact on specific places. Furuseth (1997) provides a case study of the impact of industrial livestock farming on the state of North Carolina. By 1995, pig production had passed poultry and tobacco as the leading sources of farm income, earning the state the moniker "Porkopolis." In the post-WWII period pig farming was the least integrated system in the state, and in the 1970s farms selling hogs were very small (about 150 animals per farm). In 1982, "there were nearly 330,000 farms producing hogs and pigs, but over the next ten years the number of hog farmers declined by 42% while the number of hogs sold grew by 17%" (394), with the result that between 1990 and 1995 the number of pigs increased from 2.8 million to 8.3 million. Furuseth documents how these changes came about via government and industry support and argues that the most important change associated with the industrial production model is scale. Not only did the number of animals increase exponentially because of CAFOs, where the pigs are housed in high-density cages inside buildings for the duration of their lives (several months for meat pigs, a couple of years for breeding sows), but production specialization meant that a three-tier system developed spatially. One facility handles breeding sows, another—nursery houses—takes care of the weaned piglets, and finally the pigs reach slaughter weight in grow-out buildings. This division of labor helps the scale of production by concentrating the animals at each step. Finally, the independent hog farmers who historically raised animals from birth to slaughter were replaced by contract growers that each work for large companies on one tier of the system. While business and local governments have touted the benefits of this system for North Carolina, not everyone is happy. Independent farmers have lost their farms or been forced into contract work for little money, and life next to a hog CAFO is not pleasant. Hog waste is kept in open-air lagoons that not only smell unpleasant but can rupture, sending waste into local water systems. While this push to industrialized animal production has reduced the price of pig meat, it has come at a cost to local landscapes, local livelihoods, and animal quality of life.

THE PIG: *SUS SCROFA DOMESTICUS*

The domestic pig is part of the taxonomic family Suidae, which includes thirteen species in five genera: pigs, hogs, and boars (seven species), warthogs (two species), red river hogs and bushpigs (two species), babirusas, and the giant forest hogs. Of the seven Sus species, four are considered vulnerable or endangered. All pig species evolved in the Eastern Hemisphere. The peccary, a piglike animal but in its own family Tayassuidae, evolved in the Western Hemisphere. Pigs are part of the order Artiodactyla, which means they are even-toed ungulates. Artiodactyls first appeared in the early Eocene around fifty-four million years ago, and pigs first appeared during the late Eocene around forty-six million years ago.

Wild pigs on average weigh between twenty pounds and six hundred pounds depending on species and live anywhere from fifteen to twenty years. As omnivores, they can survive on a wide variety of food sources. While some species are nearly hairless, most are covered with coarse bristles instead of soft fur. Their sense of smell, their hearing, and their vocalizations are keen. Pigs can make all sorts of different sounds, including squeaks, squeals, and chirps as well as a spectrum of grunts. Pigs use their snouts for digging and extracting food while their teeth can be used to chew, tear, or fight. Socially, most males will be mainly solitary while females form sounders with adults, juveniles, and babies. The gestation period for pigs is between 100 and 175 days and sows give birth to one to twelve piglets. They are known not so much for their graceful appearance but for their intelligence and adaptability. They enjoy wallowing in mud to cool down and remove parasites; otherwise they are fastidiously clean animals.

Wild pigs were first domesticated between eight thousand to eleven thousand years ago simultaneously in Europe, India, China, and Southeast Asia. Domesticated pigs were not used as working animals, but for food and tools (made from their skin and bones). All domestic pigs have curly tails unlike their straight-tailed wild relatives. Hundreds of breeds of domestic pigs have been developed, including the Meishan pig from China with large, drooping ears and wrinkly black skin, the Essex pig from England who looks somewhat like an Oreo cookie with a black head and hindquarters and white center, and the American Yorkshire, which has become the US CAFO standard. Pigs do have sweat glands, just very few of them, making temperature regulation of their bodies difficult. In many cultures pigs have been revered in spiritual practices, and even in popular culture pigs have made their mark, with Porky Pig, Babe the Pig, Wilbur the Pig from *Charlotte's Web*, Miss Piggy, and Piglet among the most well known. Potbellied pigs have also become novelty pets in places like the United States where they live inside and enjoy the good pet life. Pigs are also used in medicine to provide organs to humans.

Following Watts (2000) and his analysis of the enclosure of livestock into industrial systems for economic gain, Daniel Knudsen and Frank Hansen (2008) show that even farming cooperatives that are collectively owned behave similarly to larger corporate conglomerates when making decisions about restructuring. However, this similarity doesn't have to be negative. In a case study of Danish cooperatives in the post-WWII period, Knudsen and Hansen find that while the reasons for restructuring (reduced costs and increased profits) are the same, the results are qualitatively different. Denmark has a long history of livestock production via cooperatives and even today is one of the top pork exporters. In the post-WWII period, however, restructuring involved consolidating operations with the result that the number of pig farms declined from 130,098 to 11,747 between 1968 and 2002 while the number of pigs increased from 8.4 million in 1968 to 13 million in 2002. The Danish case is unique because the country managed to consolidate production systems, produce more pork than before, and still maintain wage levels for workers. Dairy cooperatives in India, however, have undergone a different transformation. Research by Pratyusha Basu and Jayajit Chakraborty (2008) on household characteristics that influence membership in cooperatives reveals discrepancies between larger development goals and local cultures. "Livestock based livelihoods are currently being promoted by international development agencies as part of global efforts to combat poverty" (300). Rural dairy cooperatives have been the focus of national attention in India since the mid-1960s, and at both the national and international scale the use of cooperatives to spur economic growth has been applied the same across the board. This generalized application causes problems because it ignores local differences over the breeds of animals and the labor that goes into dairy farming.

Livestock are not only "out there" somewhere beyond the urban areas. In many cases we see livestock in the midst of the city. Whether we are discussing the rise of backyard chicken keeping or larger-scale chicken farming, the results are the same: what the "farm" means today is changing in terms of place. Alice Hovorka's case study of chickens in Botswana "demonstrate[s] how cities are inextricably wrapped up in human-animal relations" (2008, 95). She argues that livestock in urban areas in the developing world have often been overlooked, but that they are, in fact, central to the form and function of African cities. Using the city of Gaborone as her study site she finds six key insights. The first is that urban chickens actually make up a significant part of the urban population—whether they are easily visible or not. The second and third are that, as it turns out, livestock do well in African cities as they provide ecological services such as trash cleanup—via free-ranging animals—that are not otherwise in place. This scrounging means that the economic costs to keep these animals are less because people do not to pay for 100 percent of their feed. Fourth, chicken keeping in Gaborone has actually helped the local economy—providing both small-scale empowerment at the household level and larger economic benefits to local companies. Fifth, chickens are interwoven with urban social networks that link families to communities, and finally, urban chickens are transgressing the Western urban notion that deems them "out of place." Her case study provides an excellent example of how "the city is neither purely social nor natural, but rather is produced by socio-ecological processes that become embodied in city life" (97). In essence, the chickens have become just as important a part of city life as the humans, and she demonstrates how chickens are significant actors in their own right, affecting the culture, economy, and daily life spaces of Gaborone.

While individual places change with livestock industries, as part of the global economic dynamic, places are often pitted against each other. The case of farmed catfish highlights this point. Ben Belton, David C. Little, and Le Xuan Sinh (2011) document the localized changes that have come about in Vietnam's Mekong Delta, which has seen an unprecedented rise of export-oriented Pangasiid catfish since 1997. In fact, during the first decade of the twenty-first century, Vietnam rose to become the third-largest producer of aquatic foods in the world. This boom has resulted in the solidification of class lines in the area, but not substantial class differentiation. The authors are

concerned that the consolidation of the industry locally may increase disparity in the future: not only potential economic disparity in Vietnam itself, but
a perceived global inequity between Vietnam and the United States. Becky
Mansfield (2003) shows how, as Vietnam increased its production, US producers worked to frame Vietnamese catfish as coming from inferior deltas
(i.e., dirty water) and not truly being catfish at all because American and
Vietnamese catfish are technically different species. US producers lobbied
for a labeling law, which passed, but the case speaks to the ways places and
livestock animals are constructed for economic gain. The economic geographies of livestock production then, as geographers have thus far explored,
reveal the complexity of global systems that are working at the intersection
of natures, cultures, and human appetites.

THE CULTURAL LANDSCAPE

Geographers have done quite a bit of work exploring the ways in which the
human-livestock relationship manifests itself in and on our cultural landscape. We can loosely categorize this work into three areas: specific places
where the human-livestock interaction occurs, constructions and evolutions
of landscapes, and cultural identity conflicts. Agriculture shows and hobby
farms are examples of two places where human-livestock interaction occurs.
Kay Anderson focuses on Sydney, Australia's Royal Agricultural Society
Show to show how these events enact "in thoroughly ritualistic fashion a tri-
umphal narrative of human ingenuity over the nonhuman world" (2003, 423).
In essence, she claims that "cultivation is scripted as the turning point that
launched humanity on its diverse 'civilizing' path" (423). In asking the reader
to consider the ways in which these shows reveal an obviously anthropocentric attitude, she also highlights how humans seem to need these displays of
their own prowess to confirm their place in the world. The Sydney shows are
particularly fascinating because they began at a time when European Australians were removing Aboriginal peoples from their lands in a way similar
to the removal and enclosure of native peoples in the United States. For the
European Australians, the Aborigines were a conundrum. Humans were supposed to be "civilized" and demonstrate their civilization through their ability
to control the natural world; however, the Aborigines had not domesticated
any plant or animal species and were, therefore, uncivilized. Aborigines were
thus seen as the extreme limit of the human, and the agricultural shows were
used as an "anxious reassertion of a humanity/nature divide" and a "key event
through which a European-derived discourse of civilization was constituted
and performed in the artefacts of nature's improvement" (428).

Richard Yarwood, Matthew Tonts, and Roy Jones (2010) investigate the changing entries of livestock at competitions at the Perth Royal Show over the course of the twentieth century to study the changing discourses of farming in Australia. They point out that no form of pastoral or grazing farming occurred in Australia prior to 1788 because the Aborigines never domesticated any food animals, yet in just over two hundred years Australia has become the world's largest exporter of beef, wool, mutton, and goat meat as well as the second-largest exporter of lambs and third-largest of dairy. They argue that agricultural shows are revealing for how "specific breeds of livestock embody good farming practice and technological advance in agriculture" (238). Indeed, the space of the showground "allows farmers and farm organizations the opportunity not just to display farm produce but to show different, sometimes competing, representations of rurality" (239). They document how in the nineteenth century the shows and the represented breeds highlighted sturdy stock that could be used for subsistence and local income. In the 1920s the focus shifted to breeds that had increased productivity, and the shows highlighted these genetic innovations. In the post–World War II environment, the introduction of industrial agriculture came into play, emphasizing breeding for intensive farming and specialization (e.g., lean meat). In fact, not just farmers got involved with this changing notion of the rural and livestock, but the government and increasingly multinational corporations also fully supported the shift from a more local and small-scale rural landscape to an industrial rural landscape. The agricultural show then becomes a key place in which we can see how farming—and the rural landscape—has become an increasingly complex, hybrid network connecting nature (livestock), culture (farming/rural), and technology (industrial techniques for breeding, housing, slaughtering) locally to the global scale.

While corporations farm animals at the industrial scale, and individuals and small businesses at a smaller scale, or for subsistence, Lewis Holloway (2001) points out that people are also "hobby farmers," raising animals for the pleasure of doing so—and because they are economically able to. In his case study of interviews with hobby farmers in the United Kingdom, he argues that hobby-farming is a particular type of agricultural practice in which "farm animals are encountered simultaneously as 'friends' and sources of food" (293). In addition, "human-animal relations, and the meaningfulness ascribed to them, can be regarded as significant in the constitution of place itself, as well as of human-animal identity" (294). His interviews reveal the ways in which the hobby farmers contrast themselves with commercial farmers. They believe they know the animals better because they spend more time with them and see them as "friends"; they feel they recognize that animals have their own agency and can "act back" toward humans and, in the case of the hobby farmers, are sometimes permitted to do so; and they claim the animals also help them feel a deeper affinity for their land and the relationships to their local environment. However, the slaughter and subsequent consumption of their "friends" also reveal the ethical ambiguities of this relationship to be as difficult to maneuver as they are for the retired commercial farmers.

Affiliation with livestock can get even more personal, as Mark Riley (2011) demonstrates in his case study of the emotional geographies of farmers as they retire and have to sell off their livestock. Through in-depth interviews he finds that discussing animals post-separation allows farmers to reflect more deeply on their relationships with their animals. He finds that "families of cows in these contextualized networks serve the affective function of connecting farmers to place and history" (22). In a very real sense the farmers recognize that their identities are and were bound up with the cattle themselves, and together, in the place of their farms, they became a more cohesive whole. While "culling and selling cattle is a normal, and arguably essential, part of the everyday lifescape of the farm, . . . retirement brought an inversion to the accepted sets of relations and discourses, which enable farmers to overcome the 'ethical ambiguity' that allows them to simultaneously consider cows as something 'to be cared for' or 'to be culled'" (20).

Rurality itself is being contested in terms of both farmers' and nonfarmers' constructing what they perceive as a traditional farmed rural landscape. Nick Evans and Richard Yarwood (1995) show how livestock breeds are essential in the creation of localized and unique rural landscapes. In pointing out that "anxiety has been expressed that urban landscapes are becoming increasingly homogenous, yet this does not seem to extend to rural 'cattlescapes'" (144), they argue that breeds are, in fact, essential to local landscapes. Cattle are often associated with particular places. For example, the highland cattle come from the Scottish highlands, and the Welsh blacks hail from northern Wales. The history of cattle in Scotland and Great Britain dates back to their introduction by the Romans in 43 CE after their initial domestication in the Middle East around 6000–5000 BCE. The animals themselves play a role in the creation of what the authors term "local coherences"—without which these places would not be the same. So, if we are to consider what "traditional" or more localized and unique rural landscapes look like or could be, we must consider the livestock breeds that constitute them as part of the configuration.

In a second article along the same vein, Yarwood, Evans, and Higginbottom (1997) examine the livestock indigenous to Ireland. "Traditionally, explanations for the location of livestock have centered around a breed's ability to thrive in local climatic conditions" (26); however, in today's climate, environmental influences don't tell the whole story. The authors go on to point out that the revival of Irish rare breeds can be linked to an interest in local heritage, regardless of the productivity of heritage breeds in relation to more "modern" or differently bred animals. In Ireland these breeds are being revalued for their historical qualities and their intimate connections to place and farming heritage rather than being viewed as purely an economic commodity. "At the very least, there is a need to recognize that farm animals, have, quite literally, been constructed by people to fit into particular rural spaces" (Yarwood and Evans 2000, 99). Farm animals are important within the geographic imagination of the countryside and the historical geographies of livestock breeds: where they first appeared and how/if they spread reflect the changing geographies of dominant societies and dominant imaginations in particular places at particular times. Furthermore, the survival of specific breeds depends on a complex combination of productivity, profitability, *and* local identities. What we see happening today in rural areas is a refocusing of efforts on heritage breeds and place-based livestock because many rural areas are changing from places of production to places of consumption. As the forces of industrialized farming consolidate animal production into smaller spaces, local farmers are adapting by using their rural heritage to sell the "rural" in the form of farm parks—where tourists visit to experience a working farm, consume local foods, learn traditional subsistence skills, and escape their industrial, consumptive lives in urban areas. Just as globalization has homogenized urban areas, industrial agriculture has homogenized livestock breeds and therefore contributed to the loss of localized place identities.

The close relationship between humans and livestock and landscape is not limited to more sedentary farmed breeds. Hayden Lorimer (2006) explores the connections between the three with the Cairngorm reindeer herd in Scotland. In tracing the Cairngorm herd beginning with its introduction in 1952 through time, Lorimer provides an immersive experience of "vital geographies." "In the conjoined, sinewy lives of humans and reindeer we find other matter, other properties, and other forces drawn into the realm of the 'the social.' What wells up is a biotic account of the herd enrolling winds, stones, tors [rock outcrops], trees, and mosses into a territory of patterned ground" (516). Thus, farmers and livestock exist not only on the landscape but within it. This type of connection to the environment is not possible in CAFO systems where animals have little, if no, room to experience the out-of-doors.

Building on the notion of the ideal rural livestock landscape, a new trend in livestock production has been the grass-fed livestock movement. From cattle to pigs to chickens, the idea that is farm animals are healthier and taste better if they are not given only corn to eat and have access to the outdoors. Riely (2011) highlights how grass-fed cattle in the United States numbered only fifty thousand to one hundred thousand in 2008 and how most of the farmers participating in the grass-fed trend are more highly educated and need to have their farms closer to consumers interested in these trends. This type of farming is not for high profit or intense yield, and in fact, the American Grassfed Association has very high standards for their animals and how they should be raised.

A consideration of the ways in which cultural identity shapes human-livestock relations is also a key part of understanding our cultural landscapes. All of these production systems are entwined with cultural and political identities (Robbins 1998). For example, Alice Dawson points out how similar pigs are to cattle in terms of being at "the interface of the cultural and biophysical environments" (1999, 199). Jews and Muslims see pigs as unclean and will not eat them, yet pigs were seen as divine prior to the Middle Ages and even burned at the stake with witches during the witch-hunting period. Wild pigs have been constructed as noble and full of courage while domesticated pigs have been constructed as gluttonous and full of sloth. For most of the human history that has included domesticated pigs, we have shared our home spaces and our food with them. Yet, today, pigs do not for the most part get to be seen as noble or courageous, but simply as food, and this placement of pigs into the food category has allowed us to treat them in ways that Western society does not tolerate when it comes to other species. She argues that "an important consequence of creating a dependent species is that our responsibilities and obligations to these creatures are far greater" (201), yet today we are uncomfortable with the ways in which we raise, treat, and slaughter pigs in industrial settings—especially given that animal behaviorists have determined that pigs are more intelligent than dogs—and instead choose to hide these places (out of sight, out of smell) as far as possible from consumption centers. So, in essence, Dawson is saying that we want to have our pig and eat it, too. We just do not want to have to confront that we are actually doing so and the manner in which it happens.

For most Westerners, using cattle as a food source is perfectly normal, but, like lions, other species are seen as off limits for human consumption. One of those animals is the dog. Marcie Griffith, Jennifer Wolch, and Unna Lassiter (2002) use a case study of Filipinos in Los Angeles to reveal the ways in which cultural identities intersect with place to produce social conflict. The use of dogs as a source of food has been documented throughout history and around the world; however, in today's America the practice of dog-eating is seen as out of place. Filipinos who engage in the practice are ostracized from the white community, and even Filipinos who do not engage in the practice may still be implicated by virtue of their ethnic identity. As we saw in the case of India, practices on animals as food sources have been constructed as both sites of resistance to dominant or colonizing forces and sites of separation and justification for prejudice and exclusion. We can understand racialization as "the act of classifying a group of people by assigning them real or imagined biological or cultural characteristics that subsequently are used to justify mistreatment or exclusion from mainstream society" (223). In this case mainstream society was white America.

Even in countries like Malaysia, livestock are being used in what Harvey Neo (2009, 2011) calls beastly racialization—using animals to "other" and distance social groups based on ethnic or even national identities. The Malaysian case stands out because Malaysia is a majority-Muslim country, yet pig farming is the second-biggest livestock industry after chickens. The Muslim aversion to eating pigs exists here, but the large population of pig-eating Chinese in the country and the opportunity to export pigs to the vast Chinese market have given rise to some difficult cultural-political conflicts. The rise in farming has also occurred at the same time as a growing religious sensibility within Islam. In trying to build a multiethnic country that has a history of colonization, who is a "real" Malay is often contested, and pig farming has been one way to ostracize ethnic Chinese. The local and federal governments have tried to regulate the pig industry to smooth social relations in the country, but it has been difficult to do so because the sociocultural divisions are deeply imbedded.

ETHICAL/POLITICAL GEOGRAPHIES

We have already touched on a variety of ethical and political issues associated with farmed animals, but in this section we will focus on two current topics—animal welfare and the environmental impact of CAFO production and consumption. The choice to eat animals or not is one that has plagued individuals throughout history. The Greek mathematician Pythagoras was a devout vegetarian, and for quite some time all vegetarians were called Pythagoreans. Today quite a few celebrities from Pamela Anderson to Andre 3000 of the band Outkast also choose to not eat animals. For many people the treatment of these animals is of paramount importance: they regard eating other beings as de facto unethical because they reject what they see as cruel treatment of animals. Indeed, Henry Buller and Carol Morris point out that "the mistreatment of farm animals is increasingly being evoked as a new kind of obscenity: an increasingly morally repugnant exploitation of our sentient planetary confreres which inflicts suffering in the name of profit, and in doing so, denies them intrinsic rights and identities" (2003, 216). The challenge to the hegemonic view that animals are there to be eaten has called into question the ways in which humans construct animals as "others." The problem is that "while post-modernity has encouraged us to see the individuality and subjectivity of non-humans as beings, modernity continues to put them on our plate as meat" (217).

The predominant process that puts them on our plates, for those of us used to imagining small-scale family farms where chickens, pigs, and cows wander from trough to water to sunny spot, is not very appetizing upon closer inspection. For example, typically egg-laying hens are confined in CAFO buildings at concentrations of up to 250,000 birds in a building. Anywhere from six to eleven hens are held in battery cages stacked on top of each other. Each bird has a space about the size of a regular notebook page to maneuver in for the two to three years she is there. The beaks are cut off to prevent cannibalization, and the manure runs through the cages and sits on the bottom of the building. These hens have no access to sunlight, fresh air, the ground, bedding material, roosts, or natural diets. Eggs fall through the bars onto conveyor belts where they are carried to egg-packaging machines. Breeding sows are confined for nearly their entire lives to gestation crates so small that the sows cannot turn around or fully stretch out. All they can do is lie down on their sides so their babies can suckle. They have no access to fresh air or sunlight and spend their lives on concrete slatted floors. Beef cattle, while some spend their early days in pastures, normally end up in feedlots for the last few months before slaughter at around fourteen to sixteen months old. In feedlots the animals stand in high concentrations in fenced areas in their own excrement. Food troughs are located along the edges of the fences making it easy for the animals to eat, but they have little room to exercise so they gain weight for slaughter more quickly. In the larger feedlots up to one hundred thousand cattle may stay like this for months until slaughter. For farm animal advocates these industrial methods are more akin to a factory than a farm and they usually call CAFOs factory farms to challenge the rather bland industry term.

In documenting the rise of the animal welfare movement, which seeks not to end the use of animals as food products but to ameliorate their living and dying conditions, Buller and Morris (2003) show how new configurations of society–farm animal relations are emerging. The movement argues that healthy animals equal healthy people, and the intercorporeality of animals and their human consumers calls for a rather anthropocentric reasoning for humane treatment, but one that in the time of mad cow disease, avian flu, and E. coli/ salmonella contamination resonates. Another less anthropocentric argument pushes for the subjective experiences of the farm animals themselves. The so-called five freedoms—freedom from hunger and thirst; freedom from discomfort; freedom from pain, injury, or disease; freedom to express normal behaviors; and freedom from fear and distress—highlight how the "repertoire of suffering indicators has grown rapidly both with the proliferation of industrial husbandry and with emerging societal concern to include pathological, behavioral, and physiological criteria" (229) from the farm to slaughter. A growing number of countries and companies are responding to this social shift and either legislating humane treatment or ending practices like battery cages for chickens and farrowing crates for pigs.

In considering the ways in which we can think through human-nonhuman relationships that involve the intervention of humans on the bodies and in the lives of farm animals, Lewis Holloway and colleagues (2009) use French social theorist Michel Foucault's concept of biopower—the notion that those with power not only control what/who lives or dies but also decide exactly how lives will/can be lived. In highlighting the complex array of techniques used to control the development of the bodies of livestock—by visual evaluation, by pedigree records, by breed society standards, by artificial insemination, by embryo transfer, and by genetic marker technologies—the authors argue that there are multiple sites of creating livestock bodies today, and these different networks of biopower are merging and shifting with each other, possibly without regard for the subjectivities of the animals themselves. So, in order to access—or at least attempt to access—the subjectivities of animals in these new industrial farming regimes, Lewis Holloway (2007) focuses on dairy cattle and a new technology apparatus called the automatic milking system (AMS). Holloway's work here is key for animal geography because he argues that livestock animals have been conceived of in three ways: as objects inside the animal-industrial complex, through their symbolic value to people (as rare breeds, on the landscape, in religion), or as some combination of these two. "What is so far lacking in the accounts is a more detailed consideration of what farmed animal subjectivity might mean in particular contexts" (1044). The robotic milking systems or automatic milking systems look like giant crates. The animal walks in one end, a monitor reads her tag, automatic arms come out and milk her, medicines, supplements, and food are automatically rationed out, and then the cow is allowed to leave via a door at the other end. The system is designed so that no human has to milk or otherwise touch the cows, thereby reducing labor costs and time. This "application of precision technology" means that many herds that are milked by the AMS experience zero-grazing as the animals are simply kept in pens close to the machines, and their historical twice-daily milking is radically altered because the animals are only fed if they go through the machine. The freedom for the cows to choose when they are milked is seen as an improvement in animal welfare, but as Holloway points out this may not really be the case. As it turns out, individual cows have different responses to the machines. Some cows are literally not physically able to fit the machines because perhaps their udders hang too high or too low, they may not like entering the cage, or they may be bullied by more dominant cows and unable to enter the chamber (cows live in hierarchical family groups). If the cows do not "fit" the machines they are removed, meaning farmers are now interested in breeding cows specifically to fit the machines. The lack of walking also harms the cows. With robotic systems and an extremely sedentary existence the cows are much more prone to lameness and other health problems that come with standing on concrete floors all day. In the end, Holloway wants us to consider not only that ethical issues complicate the use of these new systems but that "bovine subjectivity has a history rather than an essence and bovine being and bodily capacities are relational in terms of different technologies, economies, and social relations (with humans and other cows) that cows are associated with" (1055).

Lewis Holloway (2005) considers the bodies of beef cattle as they are configured within the shifting political processes of bodily evaluation—either visual appreciation or statistical and genetic forms of appreciation. For Holloway, "modes of explaining organisms founded on understandings of the genetic have become increasingly prevalent and during the 20th century living bodies have become reconfigured as expressions of genetic codes" (883). Animals become constituted as objects by the specific ways in which they are understood and classified. Traditionally, agricultural shows and visual methods of appreciation have been used to classify livestock, and subsequent breeding practices, as superior or not. "Through such vernacular aesthetic knowledges, breeders negotiate between understandings of commercial considerations and aesthetic evaluations in ways which reproduce specific notions of what is a 'good' animal" (890). With the increasing practice of statistical estimates of genetics, the visual appreciation of the animal no longer counts. The raw genetic data are recorded on paper and stored electronically. While both forms exist to assess individual animals relative to others, with the focus on genetics and statistics, particular types of knowledge about the animal body might be gained, but the totality of the animal is lost. The politics here, then, is not only about what type of assessment system of evaluating livestock is best, but also about the ways in which our modes of assessment render an individual animal visible as a whole entity or not. "Animals are central not only to social constructs of rurality but also to the discourses and practices deployed in political contests between constructs" (Woods 1998, 1221).

In the discourses around the mad cow disease crisis in the United Kingdom, Woods points out that cattle were represented in three ways: as biological organisms by the scientific community, as economic units by farmers, and as emblems of the rural way of life by farmers and local community members. Yet, although these animals were the focal point of contentious politics about human health, economic livelihoods, and government intervention, Woods highlights that this use of animals (perhaps like their use in agriculture itself) is always done on human terms, to further human interests and never (or very rarely) with the intention to address the needs of the animals themselves. In essence, what an animal "is" or what counts as the most relevant information to understand an animal is changing as new institutional forms of data management merge and replace the traditional expert knowledges of visual appreciation.

Jim Wescoat (1998) examines the right of animals to water in Islamic law and in the United States. He finds that Islamic law draws from three traditions: the Koran, the hadiths or stories of the Prophet Muhammad's life, and the history of existing Islamic law. He finds that animals are "seen" in all three cases but they are consistently constructed as subordinate to humans; however, animals *do* have the right to water in Islamic law. In practice, however, he finds in Pakistan that this ideal is largely not followed. When it comes to the United States he finds that "western water law does not acknowledge that animals have any inherent rights to water at all, other than what society chooses to provide in the interest of animal owners and public morals" (273). This case study blurs the line between politics and ethics, yet is key because it shows how different cultures can and do respond to the needs of animals in different ways.

Whether or not we can get people to think *and act* more ethically about food choices is addressed by Mara Miele and Adrian Evans (2010) in their exploration of whether food labels are really used by consumers to make decisions about consumption practices with meat and other animal products. Consumers have the power of choice in the supermarket even though the supermarket itself is a space designed purely for consumption. The idea of "voting" at the supermarket is becoming much more widespread and acceptable as more citizen consumers are demanding transparency in the products they are spending their money on. The main way in which we are seeing transparency is through certified labeling programs. Designating foods as organic (no pesticides, not genetically modified), free-range (animals that are not enclosed—at least for certain time periods), cage-free (chickens not kept in battery cages, but still perhaps indoors), and rBGH-free (rBGH is a genetically modified hormone used to induce cows into producing more milk) are some of the main labeling schemes right now. In using focus groups to assess whether or not labels that carry information about the lives of animals are used by consumers, they found consumers ended up being divided into two camps—enthusiastic consumers and inactive consumers. They contend that the "power to affect consumers is weaker than one might expect" (175) because consumers must work to be competent in understanding the labels and inclined to accept the responsibility of their choices. In many cases, consumers would not do either because a chore turned into something even more complicated so that consumers did not feel like they wanted to deal with the labels, nor did they feel confident differentiating competing claims.

While the treatment of livestock animals themselves can often influence people's decisions, another political/ethical issue is the links between industrial agriculture, climate change, and environmental degradation. In terms of environmental degradation, animal activists are not the only ones raising concerns. Environmentalists, scientists, doctors, and everyday citizens are becoming more concerned with the impacts of such high concentrations of animals in close quarters. Four main areas of concern have developed. The first concern is the link to climate change. The FAO of the United Nations published an eye-opening report in 2006 titled *Livestock's Long Shadow*, which instigated a global discussion of the links between climate change and CAFOs. The FAO reported that livestock are responsible for 18 percent of annual greenhouse gas (GHG) emissions, some 7.516 million metric tons globally. Although this seems like an extraordinary amount, Worldwatch Institute (Goodland and Anhang 2009), an independent research institute, published a report in 2009 criticizing the FAO for misrepresenting the actual amounts of livestock contributions to climate change. Worldwatch claims that GHG emissions from livestock are actually at least 51 percent. Respiration by livestock, a component that the FAO leaves out completely, contributes heavily to this discrepancy. According to Worldwatch, 13.7 percent of carbon dioxide (CO_2) emissions come from livestock respiration. Land used for grazing livestock as well as growing crops for feed accounts for a significant release of GHGs as well. The expansion of feed production and livestock into forested regions means that ever greater amounts of CO_2 are released. Methane produced by livestock is another human-induced GHG released into the atmosphere. Although methane does not last nearly as long in the atmosphere as CO_2, the reduction of its release could have more immediate benefits on climate. The FAO reports 3.7 percent of GHGs come from this source, while Worldwatch asserts the figure is actually 11.6 percent annually. These reports underline the fact that data may vary depending on the agency and figures used; however, whether we rely upon FAO or Worldwatch figures, the contribution to climate change by livestock is clearly a serious issue and one that is expected to grow along with the human population in coming decades.

A second area of concern is with the world's water supply. The FAO report claims that 8 percent of the global water supply is used just to water crops to feed livestock. In addition, concerns about pollution in the water supplies are increasing because of the high amounts of manure runoff, with residual pesticides and herbicides as well as antibiotics appearing in drinking water. "Huge open-air waste lagoons, often as big as several football fields, are prone to leaks and spills. In 1995 an eight-acre hog-waste lagoon in North Carolina burst, spilling 25 million gallons of manure into the New River. The spill killed about 10 million fish and closed 364,000 acres of coastal wetlands to shell fishing" (Natural Resources Defense Council [NRDC] 2011).

A third concern is impact on biodiversity. Not only have we been clearing rain forests to graze cattle and grow food for them, thereby depleting local biodiversity and contributing to global climate change, but we are also reducing the number of livestock breeds in the world, potentially paving the way for catastrophic losses if a major disease were to break out. Diversity among livestock breeds is seen as an essential tool to adapt to climate change and develop local food security.

Finally, concern is growing about the human impact of CAFO products. For example, land devoted to the production of grains for livestock feed is increasing. Globally, production of livestock feed occupies one-third of the earth's arable land. In the United States, mostly in the Midwest, 87 million acres of corn, 74.5 million acres of soybeans, and 59 million acres of alfalfa were planted in 2008, the vast majority of the resultant harvest going to feed livestock. Compare that to the fact that the top ten fresh vegetables grown in the United States come from a mere one million acres. The medical community has already been recommending we consume a smaller amount of animal products to help avoid heart disease, diabetes, obesity, and some cancers.

We can see pretty quickly how the politics of CAFOs and the environment can get contentious. Should we all be asked to stop eating animal products like we are being asked to stop driving gas-guzzling vehicles or to reduce our carbon footprints in other ways? Many people would say yes, while others would say that what they eat is their business—especially those who are in the business of CAFOs. The increasing use of humane labeling criteria, however, is demonstrating that people—in Western societies anyways—do seem to be coming to an awareness of the ways in which their animal products are being produced.

We have covered a lot of ground in this chapter in our consideration of the ways in which we farm animals around the world. For over ten thousand years livestock animals have been with us as companions and resources in landscapes, barns, fields, and factories. We have altered their bodies to fit our notions of production and tastes, and we have, perhaps, gone so far as to affect the entire planet. How we choose to use animals in farming conditions—whether for a tasty steak or a fancy fur coat—depends on our cultural location in space and time and on our ability to perceive these animals in and of themselves. Regardless of whether and how you would eat a lion, a dog, a cow, a salmon, or a pig, the fact remains that farmed animals are one of the major categories of human-animal geographies today.

DISCUSSION QUESTIONS

1. Can you be an environmentalist and consume CAFO products?
2. In a previous chapter we discussed the problem of pet overpopulation. What would you think of sending surplus dogs and cats to places around the world where they are consumed as a form of recycling?
3. Is it necessary to put faces and geographies to our food?
4. What makes it so difficult to consider widespread adoption of a plant-based diet?

KEYWORDS/CONCEPTS

aquaculture
bear bile farming
blue revolution
CAFO
domestication
factory farm

humane labeling
market/niche animal farming
racialization of livestock
subsistence animal farming
tiger farming

PRACTICING ANIMAL GEOGRAPHY

1. Go to three stores (grocery or clothing) and compare prices, labeling, and place of origin for animal products (meat, dairy, furs).
2. Research and map the farmed animal systems in your city or local area.

RESOURCES

American Livestock Breeds Conservancy: http://albc-usa.org
American Meat Institute: http://www.meatami.com
Chicken Stampede (film about the global chicken industry): http://filmakers.com/index.php?a=filmDetail&filmID=1441
Factory Farm Map (in the United States): http://www.factoryfarmmap.org
Humane Society of the United States: http://www.humanesociety.org
The Meatrix (film about AFOs): http://www.themeatrix.com
National Resources Defense Council: http://www.nrdc.org
A Peaceable Kingdom (film about animal welfare): http://www.peaceablekingdomfilm.org
Royal Society for the Prevention of Cruelty to Animals: http://www.rspca.org.uk/home

SVF Foundation for Heritage Breeds: http://svffoundation.org
United Nations Food and Agriculture Organization: http://www.fao.org
US Department of Agriculture: http://www.usda.gov/wps/portal/usda/usdahome
World Resources Institute: http://www.wri.org

REFERENCES

Anderson, Kay. 1997. "A Walk on the Wild Side: A Critical Geography of Domestication." *Progress in Human Geography* 21 (4): 463–485.
———. 2003. "White Nature: Sydney's Royal Agricultural Show in Post-humanist
Perspective." *Transactions of the Institute of British Geographers* 28:422–441.
Animals Asia. 2011. "Bear Bile Farming." Accessed August 8. http://www.animals
asia.org/index.php?UID=F81ZKLNQJPJ5.
Baldwin, James A. 1987. "Research Themes in the Cultural Geography of Domesticated Animals, 1974–1987." *Journal of Cultural Geography* 7 (2): 3–18.
Basu, Pratyusha, and Jayajit Chakraborty. 2008. "Land, Labor, and Rural Development: Analyzing Participation in India's Village Dairy Cooperatives." *The Professional Geographer* 60 (3): 299–313.
Belton, Ben, David C. Little, and Le Xuan Sinh. 2011. "The Social Relations of Catfish Production in Vietnam." *Geoforum* 42 (5): 567–577.
Buller, Henry, and Carol Morris. 2003. "Farm Animal Welfare: A New Repertoire
of Nature-Society Relations or Modernism Re-embedded?" *Sociologia Ruralis* 43
(3): 216–236.
Chaudhary, Vivek. 2002. "Visitors to Be Given Dog Meat 'to Combat Prejudice.'"
Accessed August 2, 2011. http://www.guardian.co.uk/uk/2002/may/28/worldcup
football2002.vivekchaudhary.
Davis, Diana K. 2008. "Brutes, Beasts and Empire: Veterinary Medicine and Envi-
ronmental Policy in French North Africa and British India." *Journal of Historical
Geography* 34:242–267.
Dawson, Alice. 1999. "The Problem of Pigs." In *Geography and Ethics: Journeys
in a Moral Terrain*, edited by James D. Proctor and David M. Smith, 193–205.
London: Routledge.
Emel, Jody. 1998. "Are You Man Enough, Big and Bad Enough? Wolf Eradication
in the US." In *Animal Geographies: Place, Politics, and Identity in the Nature-
Culture Borderlands,* edited by Jennifer Wolch and Jody Emel, 91–116. New
York: Verso.
Evans, Nick, and Richard Yarwood. 1995. "Livestock and Landscape." *Landscape
Research* 20 (3): 141–146.
Fields, Gary. 2003. "Communications, Innovation, and Territory: The Production
Network of Swift Meat Packing and the Creation of a National US Market." *Jour-
nal of Historical Geography* 29 (4): 599–617.
Furuseth, Owen J. 1997. "Restructuring of Hog Farming in North Carolina: Explosion
and Implosion." *Professional Geographer* 49 (4): 391–403.

Gibbs, David, Lewis Holloway, Ben Gilna, and Carol Morris. 2009. "Genetic Techniques for Livestock Breeding: Restructuring Institutional Relationships in Agriculture." *Geoforum* 40 (6): 1041–1049.

Goodland, Robert, and Jeff Anhang. 2009. "Livestock and Climate Change." *Worldwatch*, November/December, 10–19.

Griffin, Carl J. 2012. "Animal Maiming, Intimacy and the Politics of Shared Life: The Bestial and the Beastly in Eighteenth- and Early Nineteenth-Century England." *Transactions of the Institute of British Geographers* 37 (2): 301–316.

Griffith, Marcie, Jennifer Wolch, and Unna Lassiter. 2002. "Animal Practices and the Racialization of Filipinas in Los Angeles." *Society and Animals* 10 (3): 220–248.

Hobson, Kersty. 2007. "Political Animals? On Animals as Subjects in an Enlarged Human Geography." *Political Geography* 26 (3): 250–267.

Holloway, Lewis. 2001. "Pets and Protein: Placing Domestic Livestock on Hobby-Farms in England and Wales." *Journal of Rural Studies* 17:293–307.

———. 2005. "Aesthetics, Genetics, and Evaluating Animal Bodies: Locating and Displacing Cattle on Show and in Figures." *Environment and Planning D: Society and Space* 23:883–902.

———. 2007. "Subjecting Cows to Robots: Farming Technologies and the Making of Animal Subjects." *Environment and Planning D: Society and Space* 25:1041–1060.

Holloway, Lewis, Carol Morris, Ben Gilna, and David Gibbs. 2009. "Biopower, Genetics and Livestock Breeding: (Re)constituting Animal Populations and Heterogeneous Biosocial Collectivities." *Transactions of the Institute of British Geographers* 34:394–407.

Hovorka, Alice. 2008. "Transspecies Urban Theory: Chickens in an African City." *Cultural Geographies* 15:95–117.

In Defense of Animals. 2011. "Fur Facts." Accessed September 1. http://www.idausa .org/facts/furfacts.html.

Knudsen, Daniel C., and Frank Hansen. 2008. "Restructuring in Cooperatives: The Example of the Danish Pork Processing Industry, 1968–2002." *The Professional Geographer* 60 (2): 270–284.

Lorimer, Hayden. 2006. "Herding Memories of Humans and Animals." *Environment and Planning D: Society and Space* 24:497–518.

Mansfield, Becky. 2003. "From Catfish to Organic Fish: Making Distinctions about Nature as Cultural Economic Practice." *Geoforum* 34 (3): 329–342.

———. 2011. "Is Fish Health Food or Poison? Farmed Fish and the Material Production of Un/healthy Nature." *Antipode* 43 (2): 413–434.

Miele, Mara, and Adrian Evans. 2010. "When Foods Become Animals: Ruminations on Ethics and Responsibility in Care-*full* Practices of Consumption." *Ethics, Place and Environment* 13 (2): 171–190.

Nagourney, Adam. 2011. "In California, Going All Out to Bid Adieu to Foie Gras." *New York Times*, A20. Accessed October 16. http://www.nytimes.com/2011/10/16/us/ in-california-going-all-out-to-bid-adieu-to-foie-gras.html?_r=1&pagewanted=all

Natural Resources Defense Council. 2011. "Facts about Pollution from Livestock Farms." Accessed July 17. http://www.nrdc.org/water/pollution/ffarms.asp.

Neo, Harvey. 2009. "Institutions, Cultural Politics and the Destabilizing Malaysian Pig Industry." *Geoforum* 40 (2): 260–268.

———. 2011. "'They Hate Pigs, Chinese Farmers . . . Everything!' Beastly Racialization in Multiethnic Malaysia." *Antipode*. Accessed August 18. http://onlinelibrary .wiley.com.ezproxy.mnl.umkc.edu/doi/10.1111/j.1467-8330.2011.00922.x/pdf

Philo, Chris. 1998. "Animals, Geography, and the City: Notes on Inclusions and Exclusions." In *Animal Geographies: Place, Politics, and Identity in the Nature-Culture Borderlands*, edited by Jennifer Wolch and Jody Emel, 51–70. New York: Verso.

Riely, Andrew. 2011. "The Grass-Fed Cattle-Ranching Niche in Texas." *The Geographical Review* 101 (2): 261–268.

Riley, Mark. 2011. "'Letting Them Go'—Agricultural Retirement and Human-Livestock Relations." *Geoforum* 42 (1): 16–27.

Robbins, Paul. 1998. "Shrines and Butchers: Animals as Deities, Capital, and Meat in Contemporary North India." In *Animal Geographies*, edited by Jennifer Wolch and Jody Emel, 218–240. New York: Verso.

Rodrigue, Christine M. 1992. "Can Religion Account for Early Animal Domestication? A Critial Assessment of the Cultural Geographic Argument, Based on Near Eastern Archaeological Data." *Professional Geographer* 44 (4): 417–430.

Simoons, Frederick J. 1974. "Contemporary Research Themes in the Cultural Geography of Domesticated Animals." *The Geographical Review* 64 (4): 557–576.

Sinclair, Upton. 1906. *The Jungle*. New York: Doubleday.

Stassart, Pierre, and Sarah Whatmore. 2003. "Metabolising Risk: Food Scares and the Un/re-making of Belgian Beef." *Environment and Planning A* 35:449–462.

Steinfeld, Henning, Pierre Gerber, Tom Wassenaar, Vincent Castel, Mauricio Rosales, and Cees de Haan. 2006. *Livestock's Long Shadow: Environmental Issues and Options*. Rome: United Nations Food and Agriculture Organization.

Sulzberger, A. G. 2011. "States Look to Ban Efforts to Reveal Farm Animal Abuse." *New York Times*, April 13, A15. Accessed April 13. http://www.nytimes .com/2011/04/14/us/14video.html.

Ufkes, Frances M. 1998. "Building a Better Pig: Fat Profits in Lean Meat." In *Animal Geographies*, edited by Jennifer Wolch and Jody Emel, 241–255. New York: Verso.

Van Ausdal, Shawn. 2009. "Pasture, Profit, and Power: An Environmental History of Cattle Ranching in Colombia, 1850–1950." *Geoforum* 40 (5): 707–719.

Vinyard, Valerie. 2011. "Boca, Gaining and Reputation for Exotic Tacos, Plans Lion-Meat Offering." *Arizona Daily Star*, January 20. Accessed January 22. http://azstarnet.com/entertainment/dining/article_ca33ed71-b6c4-5357-8c7f -e15a91db995d.html.

Walton, John. 1984. "The Diffusion of the Improved Shorthorn Breed of Cattle in Britain during the Eighteenth and Nineteeth Centuries." *Transactions of the Institute of British Geographers* 9:22–36.

Watts, Michael. 2000. "Afterword: Enclosure." In *Animal Spaces, Beastly Places: New Geographies of Human-Animal Relations*, edited by Chris Philo and Chris Wilbert, 292–304. New York: Routledge.

Wescoat, James L. 1998. "The 'Right of Thirst' for Animals in Islamic Law: A Comparative Approach." In *Animal Geographies*, edited by Jennifer Wolch and Jody Emel, 259–279. New York: Verso.

Woods, M. 1998. "Mad Cows and Hounded Deer: Political Representations of Animals in the British Countryside." *Environment and Planning A* 30:1219–1234.

World Wildlife Fund. 2011. "Tiger Farms." Accessed August 15. http://www.world wildlife.org/what/globalmarkets/wildlifetrade/tigerfarms.html.

Yarwood, Richard, and Nick Evans. 2000. "Taking Stock of Farm Animals and Rurality." In *Animal Spaces, Beastly Places: New Geographies of Human-Animal Relations*, edited by Chris Philo and Chris Wilbert, 98–114. New York: Routledge.

Yarwood, Richard, Nick Evans, and Julie Higginbottom. 1997. "The Contemporary Geography of Indigenous Irish Livestock." *Irish Geography* 30 (1): 17–30.

Yarwood, Richard, Matthew Tonts, and Roy Jones. 2010. "The Historical Geographies of Showing Livestock: A Case Study of the Perth Royal Show, Western Australia." *Geographical Research* 48 (3): 235–248.

Zeuner, Frederick E. 1963. *A History of Domesticated Animals.* London: Hutchinson.

Chapter Six

Into the Wild: Geographies of Human-Wildlife Relations

What brings global celebrity actor Leonardo DiCaprio and Russian political leader Vladimir Putin together? Tigers. In November of 2010, DiCaprio and Putin, along with two hundred other participants, met at a conference in St. Petersburg, Russia, to discuss tiger conservation and the need for financial support. DiCaprio pledged $1 million to the World Wildlife Fund's Tiger Project, and both DiCaprio and Putin have been long-term supporters of tiger conservation efforts (Associated Press 2010). Only about thirty-two hundred wild tigers are left in the world, down 95 percent from the turn of the twentieth century, and a multicountry conservation strategy will cost upward of $300 million. Historically, hunting has contributed the most to the loss in population, but today tiger numbers are not rising despite decreases in hunting, mainly because tigers are being run out of areas where heavy deforestation is occurring, and tiger meat and parts are still wanted for traditional medicine in China. What is it about tigers that captures the hearts of people such as Putin and DiCaprio and millions of others around the world? As the largest big cat and one of the largest terrestrial predators in the world (after bears), a tiger epitomizes how complex the geography of human-animal relations can be when you are examining the category of wild animals. Tigers have long been a symbol of virility, prowess, cunning, beauty, and fear. Humans hunt them, eat them, photograph them, keep them as pets, put them in zoos, put them in circuses, and market them as cartoon characters (Tigger), stuffed animals, tattoos, and so on—all under the guise of either having a close encounter with this powerful animal or in hopes that some of the characteristics associated with the tiger will pass to the human.

Returning to the opening images of this book and the ancient cave paintings of animals, we can conclude clearly that the human-wildlife relationship has always existed. Whether we were the prey or the predator, the history of humanity could not have developed without the bodies of the wild animals we have survived on for millennia. Indeed, as humans have spread across the planet, we have had a dramatic impact on wild species. For example, human arrival and dispersal into North America contributed to the disappearance of thirty-three genera of large mammals, and after Polynesians reached Hawaii half the birds went extinct (Withgott and Brennan 2008). Today, with over seven billion humans and their over fifty billion livestock, we create enormous pressure on wild species via our consumption and their habitat loss. Paleontologists have gone so far as to say we are in the midst of the sixth great phase of species extinctions. The last phase wiped out the dinosaurs and was in all likelihood caused by climate changes from an asteroid impact; however, this time the mass extinctions are human caused. Humans may not be the only contributing factor to extinctions. The normal background rate is one species out of every one thousand disappearing every one thousand to ten thousand years. However, according to the international Millennium Ecosystem Assessment (MEA), the current global extinction rate is one hundred to one thousand times greater than the normal background rate (MEA 2005). According to the International Union for the Conservation of Nature's (IUCN) Red List, of the forty-five thousand animal species studied, 25 percent of mammal species and 44 percent of amphibians are threatened (IUCN 2011).

Do these extinctions or potential extinctions really matter? Biodiversity—the number of different species in a given area—is seen as an essential component of a healthy planet. Biodiversity confers four main benefits on the planet and thereby on humans (Withgott and Brennan 2008). First, biodiversity in given biomes provides an extensive amount of free services upon which humans depend for survival. Animals help specifically with pollination of food plants, they are themselves a source of human food, clothing, and fuel, and they contribute to localized ecosystem resilience and stability. Second, animal biodiversity enhances food security and provides potential sources of medicines. Third, animal biodiversity can contribute to local economic development through tourism and recreation. Finally, some scientists like biologist E. O. Wilson (1984) argue that humans have a need to connect with other living beings. He calls this biophilia or the "love of life." In essence, human interaction with wild animals is essential for human well-being. The stakes, then, of understanding the geographies of human-wildlife relations are quite high and speak to fundamental questions about how we see ourselves inhabiting this planet.

In this chapter we will be focusing on the geography of human-wildlife relations to see where and how geographers have explored this key umbrella category. In fact, this category of wild animals is where the bulk of the animal geography work has been thus far—not surprising given the history of geography as that of global exploration. Geographers themselves are not immune to the call of the wild, yet, as we will see, they ask us to take on new perspectives and to examine the ways in which place and space have shaped different relations in this category. While we will learn that the concept of the "wild" animal is itself a slippery one, for our purposes we can understand it as an individual member of, or group of, a particular species that has not been actively controlled by humans. While "wild" animals often live in very close proximity to humans (think of your house!), they are still not under human control. For the remainder of the chapter we will not use the scare quotes, but we will keep in mind that human-wildlife interactions are not simply about the wild out there, but also about wild animals in urban, suburban, and rural areas.

HISTORICAL GEOGRAPHIES

As we have already seen, the history of the geographic study of wildlife goes back to the first wave of animal geography, which focused on mapping species distributions and evolutionary adaptations. While this type of research continues in geography, it most often falls under the subfield of biogeography. Biogeographers researching animals today continue to make tremendous contributions to our understandings of wildlife itself by harnessing technological advances such as remote sensing, radar, banding, radio telemetry, boat tracking, and aerial surveys to help us understand more about animal lives (Gillespie 2001). Three examples from this vast literature will serve to highlight biogeographic contributions. Jenny Carter (1997) provides a solid example of biogeographic work in her study of shearwaters in Australia. In trying to assess factors that influence nest-site selection and reproductive success for these birds, she analyzes both soil and vegetation samples, finding that soil characteristics seem to contribute more toward nest selection. This ability, through biological case studies, to assess animal behavior and needs helps us understand the ways in which other species engage with their local environments and with each other—in essence giving us a deeper understanding of *their* essence. Paul Robbins, John Hintz, and Sarah A. Moore (2010) show how wolves changed the landscape when they were reintroduced to Yellowstone. Biologists saw the return of the willows because the elk population suddenly had a check. The increased number of willows allowed the reemergence of the beavers, who created wetland habitat that brought back populations of reptiles and frogs—all in all quite a rapid and immediate response to changing the configurations of animals in an area. Leonard Baer and David Butler (2000) remind us that humans are not the only ones that actively shape the landscape. Zoogeomorphology is the study of how nonhuman animals alter the landscape. For example, bears alter the landscape by excavating dens and moving earth, beavers build dams that create new wetland areas, elephants uproot trees, and birds spread plants through their droppings. Indeed, the authors state that in Glacier National Park in the United States, around 860 cubic meters of dirt is moved downslope by grizzlies excavating dens. While biogeography is not the focus of the book because it does not explicitly aim to understand human-animal relationships, biogeography *is* an essential geographic component of animal geography because it is one way of attempting to know animal others.

From a third-wave animal geography perspective, the most fundamental point to understand about human-wildlife relations is the idea that "wildlife" has meant different things to different people in different times and places. We will look at examples from the period of European colonization, local place histories, and visual media to illustrate this idea. This deconstruction of the concept of wildlife is explained by Sarah Whatmore and Lorraine Thorne when they state that "the enduring coincidence between the species and spaces of wildlife as the antipodes of human society means that, to ask what wild is, is simultaneously a question of its whereabouts" (1998, 435). What makes their point so radical is that most people today have a static concept of what wildlife is and has been and yet to "question the sanctuary of wilderness is to disturb the orthodox parameters of animal welfare and environmental concern, and to risk the wrath of those who, bolstered by scientific and/or environmental credentials, have cultivated the social authority to act as nature's interpreters and custodians" (436). Opening the Pandora's box of "wildlife," in their view, is a first step in challenging the historical separation of nature and society and thereby the dualistic separation of humans and animals. In seeking to elaborate a topology of wildlife as a "relational achievement spun between people and animals, plants and soils, documents and devices, in heterogeneous social networks" (437), they use two examples—the ancient Roman games and the monitoring of the caiman in South America. In documenting the travels of a leopard captured in Africa, transported to Rome, imprisoned beneath the Colosseum, perhaps hungry, sick, terrified, and then thrust into the arena to fight against humans or other animals to either live or die, Whatmore and Thorne argue that at this particular intersection of time and space—the early centuries of the common era in Rome—these bloody spectacles exemplified Roman notions of wildlife; no matter that the fights were highly choreographed or that a leopard would never fight a bear, or that the leopard was in a completely unfamiliar environment, for the Romans this was "wildlife." Moving to the present day and over to South America, they then focus on the management practices for caimans, who as "wild" animals are being farmed for their skins and heavily regulated. Scientific management includes giving each animal a unique numbered identity. Are caimans that are farmed, scientifically managed, and politically regulated still "wild"? According to our time and place they are. However, does the recognition of the constructions of wild animals in these ways help us to address the moral standing or these animals or even "see" them? Whatmore and Thorne conclude not really; however, the recognition of not only the typologies and networks of wildlife but also the historical and geographical constructions allows humans insight into their own cultural constructs with the potential of opening up new ways of human-wildlife being.

The history of European colonization is one not only of human-human relations but also of human-animal relations. Remember, though, that human-animal relations at this time often crossed over into human-human ones as we saw in chapters 2 and 5 as colonization was justified as a way of bringing "primitive" or more animal-like peoples into a better, more civilized state (Anderson 2000, 2003; Elder, Wolch, and Emel 1998). Focusing now on the human-animal aspects of colonization, Bernd Herrman and William Woods (2010) build on the work of Whatmore and Thorne by examining the history of sparrows and passenger pigeons in the eighteenth and nineteenth centuries. They state that "a widespread assumption exists that only historical abundance was closer to 'natural' conditions and that man's impact had only negative consequences on numbers" (176). This notion of humans being "bad" for wildlife reinforces a separation between humans and other species rather than the relational entanglements Whatmore and Thorne are asking us to see. For the passenger pigeons in the United States, their numbers were kept in check prior to the arrival of Europeans by their competition with Native Americans for tree nuts. When the Europeans arrived and began actively warring with and removing the native peoples, the passenger pigeon population drastically increased only to crash to extinction as those same Europeans who marveled at the flocks of millions of birds began killing them for food and sport. Therefore, the oft-told story of the demise of the passenger pigeon actually hides two important points—the Europeans assisted that initial rise in numbers and the population was, before that, kept in check in part through their relations with humans.

In the case of European sparrows, they argue an even more forceful case can be made for seeing them as human constructs because sparrows are strictly commensal species. As Europeans began intensifying their agricultural practices with grain production, the numbers of sparrows increased. In good years of harvest the sparrows were abundant and in bad years they were less so, but they also had a detrimental impact on food production. A European assault on raptors compounded this problem. Raptors were seen as predators that threatened small livestock such as chickens and were thus unrelentingly hunted; however, these same raptors also fed off sparrows.

On the other side of the Atlantic, Daniel Gade (2010) provides an account of the American crow. Crows fall into the category of wild organisms called synanthropes that have developed an affinity for human modifications of the landscape. He points out that with a "brain larger in proportion to its size than any other avian species, the crow's intelligence makes it much more than a creature of instinct" (152). The history of the crow before and after the arrival of Europeans is one of rising and falling populations. While living here with Native Americans, who had modified the landscape in their own way through agriculture, the crows survived quite well. As Europeans arrived and moved westward across the continent, native agriculture disintegrated and new forest environments reemerged making way for ravens instead of crows, so their numbers declined. They soon rebounded, however, as Europeans modified the landscape via deforestation and agriculture (especially corn), which reduced raven numbers and increased crows. In showing how many animals have their own "time-space tapestry," Gade concludes by arguing that synanthropic organisms, like humans, are always in the process of becoming and adapting and we do them a disservice to see them as static.

Jody Emel (1998) provides a seminal historical animal geography analysis of wolf eradication in the United States demonstrating how the construction of the wolf by humans has changed over time and how this change has had some drastically negative and positive impacts on wolves. Indeed, Emel states that "what it means to be human can never be determined without the animal other" (92) and, furthermore, argues that the construction of wolves cannot be understood without examining concurrent notions of gender, race, and economics. The history of wolves in the United States follows the story of European westward expansion. As immigrants brought from Europe an attitude of hatred toward wolves as competing predators, the animals' eradication in what was becoming the United States occurred simultaneously with the eradication of Native Americans and bison in order to make way for ranching and human settlements—the economic incentive. She points out how the period from 1875–1895 was the peak slaughter time for wolves and also the end of the bison, but not until 1914 did the federal government institute formal wolf bounties, also encouraging their death for economic gain. Who actually killed the wolves? During this period white men had primary responsibility for the wolves' demise, and Emel discusses the links between notions of masculinity and wolf killing. First, white men felt wolves had to be killed because the wolf failed to act like a proper hunter and hunted in packs, used strategy, and ate game alive—an affront to masculine ideals of hunting etiquette. Interestingly, she notes that "the construction of the wolf as a merciless killer of innocent livestock is quite interesting considering the slaughter of livestock cattle and wild bison that went on at the hands of the wolf's killers" (104). Secondly, she links wolf killing to notions of frontier masculinity where men reaffirmed their prowess, power, and domination over nature in killing wolves even as they may have admired them. This "learned capacity to cut off feelings in order to facilitate death or degradation" (106) manifested itself through intense cruelty in the killing. By using poisons, dynamite to blow up dens, and intensive slaughter, men took a nearly fanatical violent thrill in killing these animals, which marks this period of human-wolf relations as remarkable. Perhaps even more remarkably today, while many ranchers and hunters still abhor wolves, the wolves have moved in the larger cultural imagination away from being constructed as negative to becoming symbols of freedom and the epitome of the "wild."

Turning from hunting the wolves in the United States to hunting elephants in nineteenth-century Ceylon (present-day Sri Lanka), Jamie Lorimer and Sarah Whatmore (2009) explore the more-than-human history that simultaneously reveals the vital role of animals to history, deepens understandings of colonial practices, and maps complex topographies of ethical concern. Using the story of Samuel Baker they show how his hunting of elephants was linked to a specifically masculine and colonial mind-set of hot-blooded adventure. Baker considered himself a sportsman and saw hunting as civilized and necessary, but Lorimer and Whatmore point out the mismatch between the codes of sportsmanship and actual practices. First of all, the hunt was being "performed" for an audience back in Britain, not in Ceylon, and therefore stories of prowess, and the hunter subduing the exotic prey (and peoples) of a faraway land, became the goal. Hunters like Baker had to demonstrate the power of Britain and Western civilization by killing elephants. Secondly, hunters such as Baker ended up becoming spokesmen for the elephants in a strange way—by arguing that the elephant hunt was not slaughter but part of acquiring knowledge of natural history and the larger, rational pursuit of knowledge.

Other historical wildlife animal geographies focus on reintroductions and shifting human attitudes. Alec Brownlow (2000) reflects on wolves in the United States and focuses on the history of human-wolf relations in the Adirondack region of New York State. Today, conservationists have shown an interest in restoring wolves to this region, but, given the history here, he argues that the "restorationist's project of 'bringing the animals back in' presupposes necessarily an appropriate (ecological, social, political) place for animals to be brought back into . . . so is the contemporary Adirondack landscape an appropriate 'place' for wolves to be brought back into?" (141). The history of wolves in New York after European arrival occurs in two phases. In the first wolves were hunted as land was cleared for agriculture, and in the second they were seen as competitors for deer for those vacationing in the emerging playground landscape of the region. In the first wave, wolves (and Native Americans) were seen as out of place in the emerging agricultural landscape where wild animals were being replaced with domesticated ones. In the second wave the vacation economy needed the deer to be there in abundance and therefore the wolves were out of place again. In discussing the attempt to reintroduce wolves today, Brownlow argues that this full history needs to be taken into consideration because competing visions of the Adirondack region continue between insiders and outsiders who continue to frame the wolf as both in place and out of place.

A case study of wild turkeys in the state of Minnesota by Mitchell, Kimmel, and Snyders (2011) highlights the successes and failures of the history of trying to reintroduce this large bird to the area. They highlight three key points that link human and avian activity—the ability of the species to adapt to new areas, the ability to adapt to alterations of the landscape by humans, and human cultural attitudes and interventions. Reintroducing the wild turkey can never simply be about turkey biology alone, but instead must encompass an expanded understanding of both human and turkey agency. Historically, wild turkeys lived in the southeastern part of the state in hardwood forests where they had plentiful roosting locations and acorns for food. The arrival of Europeans and their agricultural methods removed the forest cover, and the newcomers also hunted them for food. Attempts at reintroduction using birds from game farms were tried in the 1920s and 1980s but with little success. The farmed birds were genetically inferior and unable to withstand the conditions of wild survival; they did not have the wild turkeys' learned history of how to survive and could not pass this knowledge down to their young. Instead, the process of trapping and transporting other turkeys already in the wild has been more successful, and today the state has a population of around seventy-five thousand turkeys. They have expanded their territory and they now survive the winters in the more northern areas because of human accommodation by doing things like not plowing under cornfields so the stalks and uncollected ears can be used as food.

After researching US federal policies to manage bird populations on the West Coast, Robert Wilson (2009) argued that World War II had uneven effects on the programs and practices of federal land management agencies. On the one hand, the US Fish and Wildlife Service (USFWS) established large refuges to lure birds away from rice crops during the war, which protected both game birds and rice. On the other hand, the USFWS employed methods of insect control pioneered by the military, including the use of DDT, in order to kill weeds on the refuges. While the new refuges restored populations of ducks, the use of DDT killed thousands of gulls and other fish-eating birds in the 1950s and 1960s. Wilson's work not only highlights the environmental effects of war, but also the complexities of wildlife land-use protection decisions.

Daniel Gade (2006) explores the human-animal interface with a case study of the historical geography of humans and spotted hyenas in the Horn of Africa. In eastern Africa today, the lion and leopard have disappeared or become so rare that the hyena occupies a novel niche and, like the coyote in the Americas or the dingo in Australia, can adapt to habitats with dense human populations. Gade finds that a close human-hyena relationship has long existed in this part of the world. Hyenas were (and still are) used as garbage and carrion removers from towns and cities because of their powerful gastric juices that enable them to digest an incredibly wide range of materials. Hyenas also prey on livestock, and Gade documents how settlement patterns historically reflect the agency of the hyena as people have constructed watchtowers and enclosures to keep humans and livestock safe. One of the most contentious conflicts, however, occurs when hyenas do not discriminate among dead bodies and choose to consume human cadavers. Historically, major famines and droughts have caused a breakdown in the social fabric that normally disposes of dead humans (via ritual burial) or even protects the living infirm, and hyenas have also taken advantage of these situations to drag humans away from dwellings and dig up corpses or even simply to drag dead ones away that have not yet been buried. For Gade, this strangely symbiotic relationship where both humans and hyenas have benefited reveals how studying the history of human-animal relations in place emphasizes how "nature" and "history" and "human" are not separate categories.

The ways in which we understand wildlife are also shaped geographically by our access to visual mediums like film, photography, and even taxidermy. In a fascinating piece, James Ryan demonstrates "how photography was used in parallel with practices of hunting and taxidermy to capture and reproduce 'wild' animals" (2000, 205). For Ryan, photography was similar to hunting because in their attempt to document living nature, photographers were re-enacting the experience of the hunt. Both taxidermy, resulting from the actual hunt, and "camera hunting" played roles in the imperial project to capture the world and express European dominance. In fact, the famous hunter/explorer Frederick Selous said that "the hunter had always been a pioneer of Empire" (quoted in Ryan 2000, 204). What were both types of hunting re-presenting to the public? Ryan argues that hunting had to do with notions of physical closeness and bravery. After all, hunting with a gun or a camera could be dangerous, and only the bravest men (at the time all men) could handle the environment. The danger was exemplified in taxidermies of lions with their mouths agape or bears reared up on their hind legs in aggressive poses. Animals could thus undergo taxidermy even if the animal that was killed was not confrontational at all! Furthermore, this spectacle served to present the "wild" as a dangerous place "out there" even as "the display of wild animals in spaces far removed from their natural habitat served simultaneously to maintain the distance between the wild and non-wild and civilized and savage" (217). As Gail Davies (1999, 2000) points out, wildlife films in the United Kingdom historically have continued this separation of humans and wild animals even as intellectual work breaking down these binaries has grown stronger. The first wildlife films came mainly from zoos and film studios, and the animals were largely under the control of humans even as they were presented as wild and dangerous or wild and meek. Advances in film technology that allowed more portable cameras and the rise of field biology in which animals were studied in situ changed the ways animals could be represented to the rest of humanity. After all, most people don't have the chance to see wild elephants, wolves, or polar bears, so we are forced to see them through the eyes of others, but these others, whether scientists or broadcasters, have shifting understandings of wildlife as well. Indeed, each wildlife film "has a distinct geography involving different configurations of film-maker, broadcaster, scientist and animal, and constitutes a particular 'culture of nature'" (Davies 2000, 453). In critically evaluating wildlife films over time and across space as boundary objects—things that set the parameters for what wildlife is (good, bad, cuddly, ferocious, out there)—Davies contends that this medium both manages to advocate for the primacy of wildlife experience while successfully framing that experience in socially constructed ways. Clearly now, while a necessary and essential connection exists between animal-focused biogeography and current animal geography, the intellectual aims complement each other, allowing for deeper understandings of both sides of the human-wildlife equation.

ECONOMIC GEOGRAPHIES

The economic geographies of human-wildlife relations revolve around two main categories of interaction: consumption (in whole or parts) and recreation/tourism. We will begin with the United States. Figures 6.1 and 6.2 reflect data compiled by the USFWS (2006) about citizen participation in wildlife hunting and watching. As you look at the maps, consider what might account for the differences among the states within the maps, and the differences between the maps. This report is compiled every five years and reflects the current state of the economics of the human-wildlife relationship within the country. The survey found that over eighty-seven million citizens over the age of sixteen participated in wildlife-related activities with thirty million fishing, twelve million hunting, and seventy-one million participating in wildlife watching/feeding.

Wildlife recreationists' avidity also is reflected in the $122.3 billion they spent in 2006 on their activities, which equated to 1 percent of the Gross Domestic Product. Of the total amount spent, $37.4 billion was trip-related, $64.1 billion was spent on equipment, and $20.7 billion was spent on other items such as licenses and land leasing and ownership. Sportspersons spent a total of $76.7 billion in 2006—$42.0 billion on fishing, $22.9 billion on hunting, and $11.7 billion on items used for both hunting and fishing. Wildlife watchers spent $45.7 billion on their activities around the home and on trips away from home. (4)

The amount of money, time, and equipment invested by people in the United States in order to have access to wildlife represents a significant economic force. In the terms of the commodity chain of hunting or wildlife watching, we want to be aware that these activities do not simply involve going out to kill a deer with a gun or capture its image with a camera, but the whole web of economic transactions that occur from transportation to lodging to food to equipment and guides. In addition, wildlife participants directly support conservation of species and places by purchasing licenses, stamps, and other

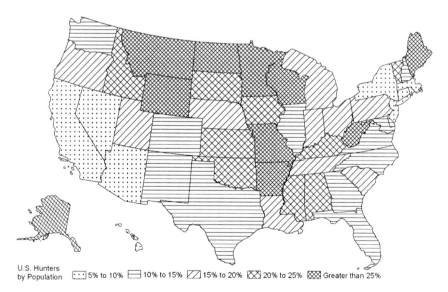

U.S. Hunters by Population   5% to 10%   10% to 15%   15% to 20%   20% to 25%   Greater than 25%

Figure 6.1.   Percentage of Hunters/Fishers in Each State. *Source:* Data compiled from the US Fish and Wildlife National Survey (USFWS 2006).

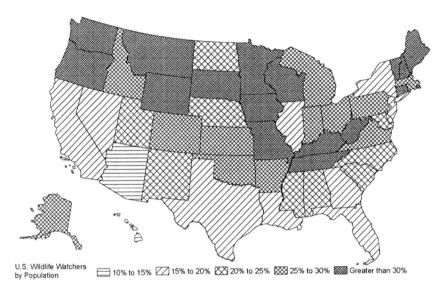

Figure 6.2. Percentage of Wildlife Watchers in Each State. *Source:* **Data compiled from the US Fish and Wildlife National Survey (USFWS 2006).** *Notes:* **Wildlife watching includes activities such as birding, visiting nature centers, photographing wildlife, and feeding wildlife in backyards.**

park use fees. While controversies do exist between people who want to hunt/ fish and people who prefer nonlethal wildlife encounters, the maps and accompanying economic statistics reveal how much US citizens invest in their own wildlife. This story is true throughout the world: the economic structure that supports lethal and nonlethal interactions with wildlife contributes to local economies and livelihoods.

While human-wildlife relations obviously impact economies positively, other interpretations ask us to be more astute in evaluating the pros and cons of wildlife economies. We can become more aware in one way by thinking critically about the relations between capitalism and wildlife. Noel Castree (1997) uses Karl Marx to do a reading of the North Pacific fur seal's near collapse at the turn of the twentieth century. Castree outlines how, for Marx, nature is not separate from society, but is materially produced through the capitalist process of creative destruction—the idea that nature is consumed in the process of transforming raw materials into objects for sale for profit. In this process the separation of nature and human society (culture) is dissolved. Furthermore, nature is also produced discursively—that is to say that "representations of nature can serve to either conceal or legitimate the often rapacious material production of nature at the hands of culture" (2).

Marx can be used to good value to analyze human-wildlife relations and moneymaking to make transparent the costs and consequences of capital's use of wildlife to turn a profit. The North Pacific fur seal was barely utilized by humans until the Russians in 1787 began killing the animals for their fur and meat on a large scale. When the United States bought Alaska, the federal government gave monopolies to two private companies, who then proceeded to overharvest the seals to take advantage of the huge demand in Europe and the eastern United States. By 1890, the seals were near collapse because the hunters killed mothers and babies at the rookery sites, and even pelagic hunting had reached a critical state by 1910. In 1911, the North Pacific Fur Seal Convention came into being as the first international agreement to conserve a species of marine wildlife. All pelagic hunting was banned and only regulated land-based hunting was allowed, which slowly allowed the seal population to recover.

While the near annihilation of the fur seal presents a classic tragedy of the commons case where a seemingly limitless resource is depleted because no business wants to limit their potential profit, Castree (1997) argues that taking a Marxian approach critical of the capitalist process reveals that environmental degradation is not only a property rights problem, but the direct outcome of capitalist processes of growth and competition where the focus is on immediate profit. Indeed, "the unregulated slaughter of seals at sea was the horrendous outcome of a mode of economic appropriation whose 'normal' functioning would have almost certainly extinguished the seal" (11). Castree also points out that with the case of representing nature, those who do the representing are also doing so for their own advantage. The discourse from environmentalists at the time created a discourse of "wild" nature and described the seals as wild, pristine, majestic animals that should not be hunted for human gain. This narrative of wild nature not only reinforces a separation of humans from nature, but also hides the treatment of native peoples—in this case the Pribilof Aleuts—by the Americans and Russians. While disentangling the economic from the political or cultural presents complications, as in this case, where does the fault of overconsumption lie? With the hunters, the companies, the government, those who incessantly demanded sealskins for fashion? Ultimately a combination of all of these led to the decline and the turnaround, but that simultaneous human construction of wild animals as wild combined with the desire to consume them beyond subsistence levels are really the bookends that frame our relations with wildlife.

In considering the role of elephant-based tourism in Thailand and Botswana, Rosaleen Duffy and Lorraine Moore (2010) bring a critique of capitalism into the present day by arguing that tourism has extended neoliberalism's reach by opening up new frontiers of nature's commodification under capitalism. Neoliberalism can be "briefly defined as a specific form of capitalism which is privatization, marketization, deregulation and various forms of re-regulation" (744). In arguing that tourism redesigns nature for shifting tastes in global consumption, they show that this catering is not necessarily always a negative thing. Tourism has been used as a means for economic advancement for countries in the developing world. However, when it comes to elephants, Thailand has a long history of training and using elephants but Botswana has none. The push to develop elephant-based tourism in Botswana is part of creating new tourist spectacles that can attract foreigners willing and able to pay. Elephants in Botswana come from circuses, zoos, and existing safari parks or were calves or young elephants who survived management culling operations. Duffy and Moore highlight how these elephants become repackaged for tourists who wish to experience the "wild" and the "exotic." While criticism has arisen that profits, not concerns about animal welfare or conservation, drive the industry in Botswana, the authors' work does show how local economies use animals in novel ways as objects of consumption.

Today, the largest international framework for trying to prevent a global tragedy of the commons for consumptive economic gain is the Convention on International Trade in Endangered Species of Wild Flora and Fauna (CITES) of 1973. The international market for wildlife goods is worth billions globally, and since trade in live animals or their parts moves around the world, an international framework is seen as necessary to provide some checks to wildlife consumption. CITES regulates over thirty thousand plants and animals and puts them into three major categories. An Appendix I listing means that a species is threatened with extinction and therefore all trade in the species is banned except in specific and exceptional circumstances. An Appendix II listing means that a species is currently not threatened but the numbers are of enough concern that trade must be closely monitored. An Appendix III listing is when a species is threatened in just one country, but member countries are asked for support in controlling trade in and out of that country.

Table 6.1 provides an overview of Appendix II listed trades in 2005 and 2010. In each case the proper import/export licenses must be obtained and data reported to CITES by member countries. The top five exporting countries in 2005 were Indonesia, Fiji, Argentina, Colombia, and Senegal while in 2010 the top countries were Indonesia, Jamaica, Madagascar, Honduras, and Colombia. Indonesia is consistently at the top of the country lists because of the vast numbers of coral and marine life being exported each year for food consumption and the global pet trade. Indonesia is also at the top because of their reptile exports. Figure 6.3 provides a visual map of these top exporting countries. We can glean from such a map that the areas of the world with the highest densities of biodiversity—the equatorial rain forest and marine environments—are the locations of the most wildlife extraction activities.

Table 6.1. CITES Appendix II Top Exports

	CITES II Top Five Exports 2005	CITES II Top Five Exports 2010
Mammals	Collared Peccary (67,530 skins)	Collared Peccary (64,980 skins)
	White-Lipped Peccary (35,500 skins)	White-Lipped Peccary (28,734
	Hamadryas Baboons (170 skins,	skins)
	8,000 live)	Vervet Monkey (5,100 live)
	Grivet (250 skins, 7,025 live)	Squirrel Monkey (3,000 live)
	Squirrel Monkey (3,200 live)	Yellow Baboon (3,000 live)
Reptiles	Tegu (1,000,000 skins)	Water Monitor (418,500 skins)
	Spectacled Caiman (602,000 skins,	Nile Monitor (262,000 skins,
	10,925 live)	11,000 live)
	Water Monitor (444,600 skins,	African Python (175,850 live)
	5,400 live)	Asiatic Python (157,500 skins,
	Nile Monitor (262,585 skins, 18,700	4,500 live)
	live)	Cobra (134,550 skins, 450 live)
	Iguana (252,200 live)	
Birds	Yellow-Fronted Canary (110,000	Orange-Winged Parrot (13,500
	live)	live)
	Red-Cheeked Cordon Bleu Finch	Senegal Parrot (13,350 live)
	(80,000 live)	Gray Parrot (9,000 live)
	Cut-Throat Finch (70,000 live)	Red-Headed Lovebird (5,000 live)
	Black-Rumped Waxbill (50,000 live)	Green-Rumped Parrotlet (4,074
	Senegal Parrot (45,300 live)	live)

Source: Data for this table comes from the CITES 2005 and 2010 Export Quota Reports.

Notes: This table shows the numbers of the top five animals being traded in 2005 and 2010 for Appendix II listed species. In most cases the skins are being used for fashion accessories like handbags and shoes while the majority of live animals are going into the pet trade. Some animals, like the caiman, may come from farms, but the vast majority of these animals are being removed from the wild (CITES data does distinguish between those caught in the wild and those born in captivity).

CITES has been seen as a tremendous success for protecting flora and fauna around the globe; however, some major concerns about consumption exist that CITES, and individual countries, have, as yet, been unable to address. The first being the illegal or black market trade in wildlife. The success of CITES depends on everyone following the rules; however, just like with drugs, guns, and the global traffic in people, the illegal trade in wildlife carries on at a tremendous rate because the economic gains are so large. Both parts of animals and live animals are illegally traded, and the global black market economy is estimated to be in the billions of dollars. Figure 6.4 provides a map of countries and the total confiscations of illegally traded animals over a thirteen-year period.

TRAFFIC is a wildlife trade monitoring network set up with the cooperation of the WWF and the IUCN. In existence since 1976, it aims to assist countries and CITES in monitoring the flow of wildlife as well as conduct research to help local communities manage their wildlife resources. Illegal wildlife is found mainly by catching smugglers, and TRAFFIC monitors

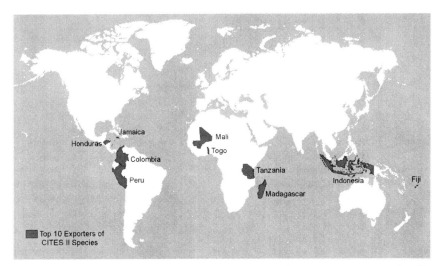

Figure 6.3.   Top CITES Appendix II Exporting Countries for 2010. *Source:* Data for this map compiled from the CITES (2011) Export Quota Reports for 2005 and 2010.

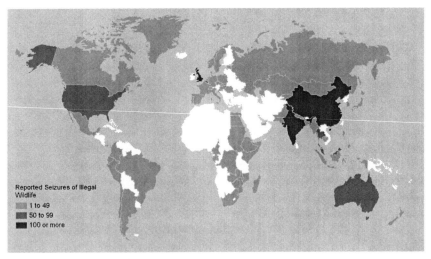

Figure 6.4.   Illegal Wildlife Seizures by Country (1997–2010). *Source:* Data compiled from the individual incident reports of the TRAFFIC organization's online news archives.

reported catches. For example, in May of 2011 a man was caught in the Bangkok, Thailand, airport with four leopard cubs, a bear cub, a baby gibbon, and a marmoset packed into his carry-on bags. He was bound for Dubai and was caught when the leopard cubs vocalized. In 2010, Thai authorities caught a woman trying to smuggle a drugged baby tiger from Thailand to Iran. The list of illegal wildlife catches is often quite bizarre with busts of people with suitcases full of snakes and spiders, making one realize how unpredictable the job of luggage inspectors can actually be. The vast majority of the illegal trade in live animals is for the pet trade while the trade in parts may be for food or lifestyle accessories such as clothes and furniture. That the trade, both legal and illegal, is so heavy speaks to the economic forces of consumption—if there was no demand there would be no consumption. While CITES is the best international framework we have right now as a bulwark against the decimation of wildlife for human desires, it is obviously not 100 percent effective.

The bushmeat trade, another aspect of the illegal wildlife trade, causes even more controversy. Bushmeat comes from the wild animals of nonindustrialized countries and is really the same thing as wild animal meat from North America or Europe in the sense that these animals are seen as local food sources. However, many of the traditional bushmeat species such as gorillas and bonobos are either listed as Appendix I or II species in CITES or considered nonfood animals by the West. The bushmeat conflict pits wealthy countries against poorer countries and often bypasses the fact that for many local peoples bushmeat is a main source of protein and something they have been relying on forever. They often cannot afford to purchase market-based animal products or do not enjoy the taste. Furthermore, utilizing local animals is seen as a local right. What would happen if the rest of the world decided to tell US hunters that they could no longer hunt certain species? The economics then of human-wildlife relations end up being much more complex once we begin to explore the specific geographies of legal and illegal trade and consumption. We have seen throughout the book that economic systems support and shape human use of animals, and the case of wildlife is no different.

THE CULTURAL LANDSCAPE

The ways in which human-animal relations manifest in the cultural landscape are as myriad as the relations themselves. For this section we will move from the intimate scale of the home to the larger scale of wilderness away from civilization. For many people their homes are a place where wild animals—insects, mice, bats, raccoons, squirrels, possums, and so on—should not be.

The home is normally constructed as a "safe" zone and exclusively for humans and their selected nonhuman companions. In a case study on brushtail possums in Sydney, Australia, Emma Power (2009) asked twenty-four different households how they deal with these animals in or near their homes. Brushtail possums, like all possums, are nocturnal animals who eat a wide variety of food. They are protected in Australia because their numbers are in decline; however, they seem to thrive in urban areas because of the easy food sources (trash) and affordable housing (human houses). She argues that human "border practices separating home from 'outside,' wildness, nature and dirt are central to the material and conceptual construction of western homes as safe, secure, autonomous human spaces" (29). What she found reveals how contingent human-wildlife relations are on an individual basis because of this construction of the home as a hard border to wildlife. On the one hand she found that several families rejected the brushtail possums and actively removed them or tried to prevent them from entering the home, reinforcing the notion of human-nature borders. On the other hand, many of the participants responded that the possums themselves actually contribute something to the feeling of home, some even going so far as to worry about the possums if they don't come "home" at night. Instead of seeing the home as a castle to be defended from the wild enemies, those that were more enamored of the possums accepted the home as a fluid space with porous boundaries, even helping these households feel more of a connection with the natural world. Furthermore, the households that accepted possums also felt a certain pride that they were doing something good for the country as well—helping protect native species. In both cases, Power argues, we can see the agency of the possums either through their ability to enter homes or their ability to (dis)enchant humans, and recognize that the process of human-wildlife relations occurs because of the actions of both humans and animals.

Understanding the role of gardens in shaping human-wildlife relations is, according to Lisa Naughton-Treves (2002), a key piece in understanding the cultural wildscape because gardens, like houses, serve as boundaries for wildlife being in or out of place. She conducted field research in Tambopata in southeastern Peru to understand how local gardeners viewed local wildlife and how this might impact local biodiversity conservation practices. She found that wildlife move easily across garden boundaries and respond more to regional conditions than individual gardener actions. In fact, "wildlife defy the boundaries humans impose on the landscape, and people have only limited ability to keep 'good' animals in and 'bad' animals out. Ironically, the very animals that inspire popular conservation campaigns are hard animals with which to live" (502). The consequences for wildlife management are important here. First of all, she argues that the garden metaphor conjures a false, ancient dream of humans' peaceful mastery over nature, and the metaphor of the garden cannot simply be expanded to a space as large as a park. Wildlife survival is shaped by what goes on beyond the borders of the garden or park.

Taking a different direction with the wildscape of gardens, Paul Cammack, Ian Convery, and Heather Prince (2011) argue that gardens are important as sites for both leisure and conservation, and they have been overlooked as such by most conservation and environmental management research. Gardens are seen as good places because they link nature and everyday life. In their study they interviewed self-identified bird-watchers in the United Kingdom and found that many altered their landscape and their habits to entice birds to come into their gardens. They might provide food, water, nesting materials, or sheltering plants and in the process of doing so form close attachments to the birds they enjoy watching. The use of gardens to promote education and conservation can be viewed as a process of collaborative interaction between humans and birds.

In Glasgow, Scotland, animal geographer Michael Campbell (2007) has studied human-bird relations via field observations of both humans and birds in the city's parks. He finds that the adaptability of each species to the myriad green and built spaces, competition between species, and tolerance for human presence have implications for a more critical urban biogeography and ecology—one that isn't exclusively about the animals alone. In arguing that conservation measures need to include the human-animal relationship as well as the nonhuman dynamic, Campbell makes the case for how seeing animals in new ways can change conservation practices. The cultural wildscape of Japan is also in flux as the urban Japanese population increasingly seeks to bring animals back (Waley 2000). While conflict has arisen about whether or not the animals belong, people also have "a sense that animals should be there—that an unanimated nature is no nature at all" (159). Waley notes that the Japanese do have a strong anthropomorphic attitude toward animals and an interest in bringing fish such as salmon back to urban rivers and fireflies back to urban landscapes, and working toward more native and freeform landscaping rather than an exhaustively tamed and idealized landscape. This newfound welcoming of nature is part of not only a changing construction of where wild animals should be but also part of a shifting paradigm of what constitutes the urban.

The boundaries between what is wild and what is domestic—or human controlled—are taken for granted according to geographer Sandie Suchet (2002). Echoing the deconstruction of the term *wild* of Whatmore and Thorne, Suchet argues that Western notions of wild management, in which tracts of land are separated from humans, create a false distinction. False because "wild management" renders invisible in the landscape all the ways in which humans make their mark on the wild—via roads, fences, created water sources (ponds/lakes), culling, and tourist infrastructure (e.g., signage, restrooms, picnic tables, lookout points, boardwalks). She points out that Aborigines have a very different concept of the wild—seeing the human-constructed landscape as degraded and wild while seeing what the Western view would describe as the wild as quiet, nondangerous, and preferable. In reminding us that the definition of "wild" is culturally and materially dependent on context, she argues that these types of universalized assumptions "can be unsettled and challenged" (153) in order to construct possible new ways of being with the "wild." David Lulka (2008) asks us to consider from a different perspective the in-between rural, urban, and wild spaces through a discussion of roadkill: "entities created through the dissolution of wilderness and the more intimate integration of society and nature" (38). Lulka points out that *roadkill* as a term "diminishes the significance of deceased animals through homogenization, as a diverse array of biological organisms is placed within one conceptual category" (39). Most people's simultaneous feelings of unease or sadness and inevitability with the various corpses of animals that line our roadways serve as literal markers on our landscapes of the disappearing wilderness at human hands. Like the Aborigines, however, he argues that we can take a new perspective and understand our responses and the dead animals as examples of a shared human and animal persistence to live, thereby giving agency to the animals and acknowledging our own discomfort with our connections to wildlife.

Building on Suchet, David Matless, Paul Merchant, and Charles Watkins (2005) use the term *animal landscapes* to describe the spaces in which human-wildlife relations become configured and contested. These spaces include committee rooms, wildlife parks, rivers, marshes, the sky, homes—all the locations where we either engage directly with what we determine to be wild animals or work to manage those very animals. In exploring the differences between otter hunting and wildfowl hunting in England between 1945 and 1970, the researchers show that in the transition to a more modern environmental awareness and sensibility, the waterfowl hunters moved with the times, while the otter hunters became cast out as archaic, even though both sports involved the killing of animals. This difference in fate came about because the waterfowl hunters shifted into a role of efficient and nonwasteful hunter that could actively abide by regulation. Furthermore, waterfowl were not seen as individuals, while the otters, because of their dramatically smaller numbers, became more individual and their slaughter seemed less justified. So, even in the same time period and relatively same place, two very different human-wildlife relations emerged. In another study of the ways in which wildlife are refractors of cultural values in the landscape, Matless

(2000) explores how the bittern (bird) and the coypu (giant ratlike creature) become implicated in what the Broadland region of the United Kingdom is "supposed" to be. The nativeness of the bittern reflects a notion of the proper English wildscape whereas the introduced coypu—an animal native to South America that was brought over for farming but then exploded in population—is seen as alien. In this human construction of properly and improperly placed wildlife, he is also concerned with how animals and humans become subjects and objects because the ability to experience animal others ends up shaping the cultural animal landscape. He says that animals can act to define humans as visceral observers where the metaphoric or literal connection between eyes, guts, and gun serve to bring the visceral human closer to the visceral animal. This relationship differs greatly from that experienced by his second category of humans—reserved watchers. In this category, even though humans are making use of their sensory organs, they are doing so in a more reserved mode and are detached rather than viscerally connected. It is not only our constructions of the animals, but our ability to experience the animal other that shapes place-based human-wildlife connections.

A wildscape that includes fish is not often studied, but Christopher Bear and Sally Eden (2011) have investigated how recreational anglers make sense of and engage with fish. In seeking to remedy an animal geography focused almost exclusively on land animals because fish are arguably difficult to study due to the spaces they inhabit and their bodily characteristics—being not as charismatic as animals such as wolves or even possums—the authors used focus groups and interviews to explore "angling as a transformative practice whereby anglers and fish adapt through their coconstitutive encounters" (336). They found four ways in which humans and fish related to each other. In the first case, the anglers got to know the landscape of the fish as they studied fish behaviors even to the point where they could recognize individual fish. Second, the anglers learned to differentiate between species in terms of which ones liked which lures, the time of day they would be most active, and where the different species hung out. The fish were also seen to respond to the human anglers by changing their behavior—shifting locations, shifting times of feeding, moving away from boats, and so on. In this way the human-fish relationship demonstrates active agency on the part of both species as they get to know each other through the barriers of water and air. Running parallel to the study of anglers and fish, Mark Bonta positions "birding as an extraordinarily intimate exploration of place, reinforced by anticipation, repetition, experience of beauty, and the culminating encounter of human self, bird, and landscape" (2010, 139). He calls this experience ornithophilia after geographer Yi-Fu Tuan's notion of topophilia or love of place. He highlights how four components help humans "become bird"—a state "wherein the birder becomes ecstatically entangled with the other and at a certain level stops 'being human'" (149). These steps include anticipation of seeing birds, repetition through seeing the same birds and learning their calls, behaviors, and colorings, enjoyment of the aesthetics of birds in flight, in display, and in plumages, and finally fulfillment of all of this happening between a human self, birds, and a specific landscape.

Tourism, like hunting, shapes human-wildlife relations in specific places. Researching brochures for safari tours in East Africa, Andrew Norton (1996) points out how tourism has become increasingly part of cultures of consumption, and the places of consumption have been manipulated through advertising. In fact, Norton claims that "tourism is one of the most important elements shaping popular consciousness of places, cultures and natures" (356). East African nature becomes marketed as a primeval landscape, reproducing a romantic discourse that places wild animals and local peoples as primitive remnants of prehistory. John Connell (2009) takes a different view of tourism through his study of birders who willingly spend large sums of money and travel to distant places to experience novel species. While places are being constructed as sites for birding, Connell argues this construction doesn't have to be a detriment and, in fact, can help local economies without a negative impact on local sustainabilities.

Paul Cloke and Harvey Perkins (2005) focus on cetacean tourism in Kaikoura, New Zealand, to explore how "the nonhuman agency of nature is implicated in the performance and meaning of place" (2005, 903). Recognizing that a triangulation of landscape, human activity, and nonhuman activity creates the tourist experience, Cloke and Perkins argue that this type of ecotourism "often represents a mix of the zoo and spectacle, experience punctuated by magical 'trophy moments' of encounter with animals 'in the wild'" (907). Yet while tourists may feel they are experiencing wild encounters, the tours are actually highly mediated processes that use technology (boats, GPS, diving equipment, cameras) to track the cetaceans and take people to them. Tourists both recognize and overlook this process because of the lure of the "place-experience of encounter with these most special of animals, and the experience-performance of getting in amongst the whales and dolphins in their own world, [and] seeing them perform their trademark manoeuvres" (911). Their work here echoes the findings of David Duffus (1996), who studied the recreational use of gray whales. Since many whale populations that were formerly hunted have now become objects for tourists to enjoy, Duffus raises concerns about the impact on the whales themselves. He finds that the whales have learned to move farther away from the tourist boats as they act to protect their own interests and life experiences.

THE WHALE AND THE DOLPHIN

Whales and dolphins are of the taxonomic order Cetacea of which there are two suborders—Odontoceti (toothed whales) and Mysticeti (baleen whales). The seven families of toothed whales include river dolphins, dolphins, porpoises, beluga and narwhals, sperm whales, pygmy sperm whales, and beaked whales. The four families of baleen whales are the gray whales, rorquals (blue, fin, humpback, and minke), right whales, and pygmy right whales. The earliest cetaceans first appeared during the middle Eocene around forty million years ago. The split between the baleen and toothed whales occurred about thirty million years ago during the Oligocene. Whales and dolphins are found throughout the world's oceans, and several species are endangered or vulnerable.

Whales and dolphins are classified as marine mammals because they are warm blooded and give birth to live young who feed off of milk. They do not have hair like land mammals so in order to keep themselves warm in the cooler water environments they insulate themselves with layers of fat. They range in size from 3.5 to 90 feet in length and from 88 pounds to 150 tons in weight. The toothed whales feed mainly on fish and squid while baleen whales have a system of horny plates rather than teeth, which they use to strain water to eat the plankton and smaller marine fish. While many species spend their lives closer to the surface of the ocean, other species dive for their food. Sperm whales have been recorded diving to depths of nearly nine thousand feet. All whale species use vocalizations, and the toothed whales use echolocation to find food. The smaller species can live for twelve to fifty years while the larger species can live up to one hundred years. Social organization depends upon the species with some living in large family groups and others spending more time alone. The dolphin family, which has evolved over the last ten million years, is known for its intelligence. Dolphins have the ability to perform complex tasks, develop abstract thoughts, and demonstrate self-awareness.

While whales and dolphins have never been domesticated, humans still have had a long and complex history with them. Records of whale hunting go back thirty-five hundred years in Alaska and around one thousand in Europe. Whales have been hunted for the oil in the blubber and their meat. Overconsumption of whales led to precipitous declines in many populations by the early twentieth century, and the International Whaling Commission was formed in 1946, but not until 1982 was a global ban on commerical whale hunting instituted. Today, only a couple of countries in the world hunt whales for either scientific purposes or as part of indigenous people's traditional cultures. People today mainly encounter free whales through in situ tourism or as captive whales in marine parks. Keeping whales and dolphins in captivity performing for humans causes controversy given how much captivity limits their experiential lives. Lolita, a killer whale who has lived a solitary existence for more than forty years in a pool in Miami, is often a rallying point for freeing captive cetaceans. The most successful release program was with Keiko, the killer whale who starred in the film *Free Willy*. Keiko was caught as a baby in 1979 and spent fifteen years in captivity in Canada and Mexico before he became a film star, and the Mexico amusement park where he lived agreed to donate him to the Free Willy Foundation. After being flown to Oregon for a couple of years he was flown in 1998 to Iceland where he experienced being in the ocean for the first time. Between 1998 and 2003, when Keiko died, he survived on his own and traveled thousands of miles in contact with other whales. He is considered one of the greatest wildlife reintroduction stories.

Indeed, Katja Neves (2010) highlights the uncomfortable fact that present-day cetourism is closer to the processes of historical whale hunting than most people probably realize. The problem for Neves is that marine ecotourism has become equated to ecologically sound conservation practices by environmentalists; instead, Neves asks us to recognize that cetourism "reflects a much wider late capitalist trend in which conservation is increasingly conflated with consumption" (721). In addition, all cetourism is not the same: it is done differently in different places, thereby impacting local whale populations differently. She shows how most cetourism does not take into account issues such as underwater noise pollution and the effect it might have on cetacean communication, or the stress an excessive number of boats can cause to family groups, especially those with young, or even the stress caused by humans constantly wanting to jump in the water and swim with them. The similarity to whale hunting is that the whales are commodities in both cases. In the first case, the literal body of the whale is extracted for profit; in the second, the living whale becomes the product being sold for profit. In both cases long-term and place-based disregard for cetacean lives is continued. Therefore, while environmentalists argue that cetourism indicates more enlightened understandings of nonhuman others, Neves argues that pointing to whales' recommodification under the guise of conservation has just as much validity.

Another interpretation of cetourism focuses on the constructions of dolphins in New Zealand in relation to sex and gender (Besio, Johnston, and Longhurst 2008). On the one hand tourists are offered a chance to experience wild nature and sexualized others, with advertisements selling the sexiness of dolphins widespread. On the other hand, the dolphins are anthropomorphized as good mothers and constructed as domestic nature. While recognizing that "the embodied experience of dolphin tourism—being in dolphins' spaces not just gazing upon them—produces an intimate connection between the seer and the seen, between humans and animals" (1222), the authors are concerned that the marketing of sexy mothers obfuscates the dolphins themselves.

An exploration of gorilla tourism near Bwindi Impenetrable National Park in Uganda by Ann Laudati shows how "'new' relations between people and parks created under ecotourism in Bwindi have in actuality created new forms of control and vulnerabilities" (2010, 726). Bwindi became a national park in 1991, and the use or extraction of any forest resources by community members was made illegal and subject to fines and imprisonment. This myth of Bwindi as an island of biodiversity surrounded by a sea of humanity plays a powerful role in foreigners' imaginations of Africa and Uganda today. However, this literal construction of a separate human and natural landscape fixes local peoples spatially and metaphorically as "noble savages," disregarding their desires to live differently or utilize their own lands as they see fit. Because foreign tourists come to not only see the gorillas in the wild, but also to see the local people living their "traditional" lifestyles, both humans and animals become frozen objects for a foreigner's gaze. Furthermore, as the gorillas become more habituated to the presence of humans through tourism, they are moving out of the park and into local communities, raiding crops and causing damage. Such "staged authenticity," for Laudati, means that "ecotourism aids in the subjugation of native people through postcolonial constructions that rely on outside images of nature which benefit nonlocal and globalized interests" (733) because it requires that "farmers remain spatially segregated from park resources so tourists can experience a perceived wild and primitive nature untouched by people" (740). What we have seen in this section is the spectrum of ways in which we encounter wildlife in our everyday landscapes.

ETHICAL/POLITICAL GEOGRAPHIES

How do our ethical obligations to wildlife play out geographically? As Steve Hinchliffe reminds us: "developing closeness to companion species, urban wilds, rivers, is also to recognize differences between and within any given setting. Living with others is partly a matter of learning to understand our co-dependences, our co-evolution, but also, to respect their differences from and indifference to us" (2007, 163). Wolch and Zhang use a survey format to assess "how demographic traits, socioeconomic status, personal background features, and past or present geographic and cultural context might shape attitudes toward marine wildlife in the Los Angeles coastal zone" (2005, 466). They find that women are more likely to support bio/ecocentric discourses about marine wildlife, women are less tolerant than men of controversial practices, and a larger differential exists between racial/ethnic attitudes than between genders with more Caucasians claiming bio/ecocentric viewpoints than any other racial/ethnic category. In El Salvador, Michael Campbell and Maria Alvarado (2011) studied public perceptions of jaguars, pumas, and coyotes. While jaguars and pumas are extinct in El Salvador, coyotes are widespread. In surveying public perceptions of reintroductions and living with wildlife, the authors found that in general people agreed that jaguars and pumas should be reintroduced to rural areas and zoos, that wildlife should be tolerated and relocated instead of shot, and that while the animals are dangerous to children they are still good for humans overall. Furthermore, younger respondents were more supportive and tolerant. These two surveys reveal that places matter when it comes to living with wildlife and also that people do reflect on the human-wildlife relationship in different ways. In this section we will explore four areas of human-wildlife relations that animal geographers have studied: conceptual critiques of wildlife, wildlife and the moral landscape, human-wildlife conflict, and the politics of engagement with wildlife.

Conceptually, animal geographers are turning a critical eye on our constructions of wildlife. Whatmore and Thorne (1998, 2000) have argued that wildlife is a "relational achievement" not only between humans and animals, but also between humans, animals, and documents, devices, and other practices that help constitute specific human-wildlife constructions in particular places, such as scientists' using tracking devices to monitor bear movements or USFWS's giving out only so many permits to shoot male elk. In their 2000 article, they outline what they call the "spatial formation of wildlife exchange (SPWE)" to "emphasize the diverse modalities and spatialities of these social mobilizations of wildlife to focus attention on the distribution of the effects and shifting positionalities of animals in and through them" (187). These social mobilizations are a political project because they shape who gets a say in where and how the image, genes, and lived lives of animals are controlled. Whatmore and Thorne demonstrate the link to the political with a two-part case study approach to the ways in which elephants are "mobilized"—first through an analysis of records of lineages and breeding and second via analysis of in situ conservation research. They then compare and contrast these two examples with "three simultaneous moments" (2000, 187) in each network: the elephants as virtual bodies, as bodies in a place, and as experiential subjects.

With elephants as virtual bodies, Whatmore and Thorne point out how using the International Species Information System (ISIS) to track elephant genetics means that animals "circulating in this fragmentary way are traceable to living creatures only by means of numeric codes tattooed or tagged on their bodies" (2000, 189). In this way the elephants are mobilized as genetic resources to be bred, moved, or otherwise controlled for the best genetic management without regard for their living selves. In the case of in situ conservation research, they explore the virtual bodies of elephants as they are displayed in the brochures for the Earthwatch citizen vacations, on which you can pay to participate in elephant research in the wild (well-protected areas). Here, elephant images that appeal to humans' appreciation of the animals as majestic combine with hints of being able to experience a firsthand and up close encounter with a wild elephant to mobilize the elephants according to a romanticized human view. In considering the mobilization of elephant bodies in places, the authors discuss how an elephant in a zoo is much different than an elephant in a reserve. "The layout and arrangement of animal enclosures in the space of the zoo are at least as forcefully shaped by the passing spectatorial sensibilities as by those of their permanent inhabitants" (191). With Earthwatch, animal bodies in place are mobilized by the opportunity to walk in the presence of elephants and to see their droppings, their footprints, and the evidence of where they've rubbed on trees or pushed them over, in the process learning to see the landscape itself as well as the elephants. Regarding the experiential lives of these animals, Whatmore and Thorne point out that Duchess, a captive elephant in Devon, England, "taxonomically . . . certainly belongs to *Loxodonta africana*, but the elephant she has become through her life at Paignton Zoo bears only distant relation to those of her kind at home in the African bush, even as such living spaces are themselves being increasingly reconfigured in the same patterning of foresight in which she is caught up" (194). In essence, despite being categorized as one, we cannot "know" Duchess as a wild elephant because she has been so effectively cut off from her "wild" existence. The wild elephants, however, still become unknowable to a certain extent because they, too, are increasingly controlled spatially and have to learn to live with human onlookers. This article is essential to a geographic way of understanding wildlife because, through their concept of SPWE, Whatmore and Thorne show how wildlife—whether an individual or a species—is constructed through a combination of human, animal, and technological networks.

Another concept often taken for granted when it comes to wildlife is that of biodiversity. Jamie Lorimer critically evaluates this concept and argues that it "must be understood as the discursive and material outcome of a socio-material assemblage of people, practices, technologies, and other non-humans" instead of as "a set of objects and processes revealed to us by an all-seeing, disembodied natural sciences" (2006, 539–540). While the concept of biodiversity was codified at the 1992 Convention on Biological Diversity at the Earth Summit in Rio de Janeiro, Brazil, as the variability among living organisms, including diversity within species, between species, and between ecosystems, in practice it has come to mean different things in different places. With species action plans (SAPs) a species has to go through (or is subjected to) four different steps—the description of the species, surveillance and research into populations, evaluation (is it threatened, endangered, or not), and finally the action plan itself. Seemingly straightforward, the actual politics and practicalities of which species get attention and which don't reveal the practice of biodiversity to be one of subjective, not objective, science. Robert Crifasi (2007) provides an example of this subjectivity with the confusion over the Preble's meadow jumping mouse in the US west. Taxonomists first classified the mouse as a new subspecies, which led to its being considered for listing as an endangered species, which would have tremendous ramifications for local development. In the subsequent conflict over the listing, taxonomists could not come to a complete consensus as to whether or not the mouse was a distinct subspecies—making the process of doing science more explicitly social and political than the normally "done deal" of taxonomic classifications.

David Lulka (2004) brings the question of ethical relationships with other species, especially wildlife, to the fore in his critique of wildlife management practices with a case study of bison at Yellowstone National Park in the United States. Current management practices focus almost exclusively on genetics and containment, which, for Lulka, are inadequate because they separate "essence from experience and facilitate the removal and exclusion of nonhumans" (439). In effect displacing them at both levels. In drawing on the work of philosopher Gilles Deleuze, Lulka is constructing an ethical argument of movement—"an ethics in which movement is viewed not only as a means of redefining human-animal relations, but also as a means of facilitating agency" (440). The bison in Yellowstone, a small remnant population from the hundreds of millions that used to live in the central plains of the United States, are heavily managed as wildlife of the park. The main problem for the bison is that they carry brucellosis, a disease that livestock cattle are also carriers of and can get. Brucellosis can cause a variety of reproductive health problems such as decreased milk production, aborted or weak calves, lameness, weight loss, and infertility. The bison must be managed so they do not come into contact with the livestock, and the best way to do that is to keep them in the park. Traditional management practices included shooting, capture and return, and culling the herd to make wandering less likely. In pointing out how the genetic management of the bison is based on the notion of a minimum variable population, "the focus on genetic variation has become the 'central dogma' of conservation theory and practice" (444); however, in practice this focus has resulted in a framework that does not see the whole bison—only its genetic material—and forces them to adapt to modern spatial structures. To counteract this confounding complicity between livestock ranchers, biologists, and environmentalists, Lulka argues that "genetic representations of nonhumans will need to be discarded in favor of more expansive approaches that foreground the embodied, visceral nature of existence and encourage fluid human-nonhuman relations" (449). Management has sought to "stabilize" the bison—genetically and physically—in place but the bison themselves have adapted to the management practices by forging rivers, using roads and trails in the parks, and continuing to not "get it" that they are supposed to stay in the park for their own good. Lulka argues that the goal of wildlife management should be to understand and facilitate biological movement between multiple species rather than containment. If these goals are not altered, then conservation ideas about genetic control and sustainability are being subsumed into more traditional dualistic conceptions of human-wildlife relations that situate animals as static masses of genetic material devoid of agency or experiential existence. How is that an ethical relationship when we know that the process of existing and becoming is applicable to both humans and animals?

Christopher Bear and Sally Eden (2008) extend Lulka's argument to the spaces of fisheries certification. In examining how the Marine Stewardship Council (MSC) constructs sustainable fisheries, Bear and Eden show how the fish themselves—because of their mobile nature—are fluid agents and not static beings whose needs and spatial movement must be taken into account. In fact, "the identity of the fish is partly the result of the regional spaces through which they travel" (501). Therefore, while the MSC can create distinct boundaries for sustainable fisheries, they only truly control the harvesting practices, not the fish themselves.

These critiques of constructions of wildlife provide a background that helps us better understand our second area of human-wildlife relations, which is the study of specific ethical/political controversies about wildlife management and the moral landscape. James Proctor (1998) explores how the controversy over the listing of the spotted owl as an endangered species veils more complex processes of culture that create what he calls moral landscapes—literal biophysical places that are imbued with notions of good and bad. While the conflict over the listing of the owl is a political one, Proctor argues that it goes beyond that to exemplify a clash of meanings about wildlife, place, and human-wildlife relations. In essence, landscapes are biophysical locations, but for humans they are also sites of emotional response—a sense of place—"to the point that differentiated human meanings become embodied in apparently objective features of nature" (192). Historically, the owl has been constructed in different ways: as an embodiment of wisdom, a humanlike animal, a killer, an omen of death, and, in the case of being listing as endangered, a symbol of primeval nature untouched by humans. He finds after reviewing the various stakeholder positions, their intent, the mechanism of their messages (TV, pamphlets, etc.), and the outcome that the animal geography of the spotted owl controversy is "shaped as much by the ideological production and consumption of moral landscapes as by the biology of the spotted owl and its habitat" (195).

A related case study of the intersection of place, science, and politics about wildlife comes from northern Spain. Ismael Vaccaro and Oriol Beltran (2009) document the conflicting constructions and politics of returning populations of wild species to the Pyrenees area. The twentieth century for this area has seen human depopulation, economic collapse, and migration of people to urban areas. For wildlife this situation offered the ability to return to areas that were previously uninhabitable because of human density and landscape use. As interest in wildlife conservation and landscape preservation has increased over the past several decades, this area has become what they refer to as a "state-sponsored zoo" where "guided environmental recovery appears to be about re-creating idealized landscapes that may not be associated with previous species in the same territory," resulting in "the cultural production of a landscape in which nature is reinvented to fulfill our postmodern standards for wilderness" (502). The role of science is supposed to be objective and turn wildlife into quantifiable components of biodiversity goals; however, species have different cultural weights, and this difference has had an effect on decision-making processes. Species could be seen as charismatic, invasive, pestilent, key, extractable, or endangered and thus managed differently even as they are all supposed to be "wild." As examples they discuss wolves, bears, beavers, otters, marmots, mouflons, and feral goats. Bears were reintroduced into the Pyrenees from Slovenia because local bears were extinct, but this reintroduction caused a negative response from local inhabitants who feared the bears. Beavers have reappeared and are successfully breeding after being gone for over three hundred years, but the push to reintroduce them came not from the state itself, but from environmental activism. The intentional release of the beavers was depicted as a way to challenge what is seen as top-down state management. The wolves have spontaneously reappeared from territories north and east. The mouflons have been forcibly removed in some areas because they compete with livestock, even as feral goats have become a nuisance and a challenge to the natural order. Therefore, far from a consistent management policy based on ecology, "the described processes depict a landscape in which factors such as taste for more biorich environments (marmots and bears), political conflicts over legitimacy (beavers and bears), attempts to re-create past environments (bears, wolves, and elk), or considerations about ecological integrity (mouflons and goats) influence 'scientific management'" (510).

Henry Buller (2004) argues that wild animals are increasingly seen as more authentic members of rural landscapes. He also studies wolves in Europe, and in his research into the reintroduction of wolves to the southern French Alps, he explores "the competing 'philosophies of nature' that are revealed when agendas of biodiversity enhancement and protection conflict with notions of biosecurity" (Buller 2008, 1583). Historically, the wolf in Europe has been seen as the über-dog and the wild and untamed. To appease humans' sense of security, wolves have been simply removed as much as possible from the landscape; however, in the second half of the twentieth century the risk of security has been supplanted by a new risk—that of extinction. In reintroducing wolves to the Mercantour National Park in France he notes that the wolves are returning to a landscape in which they are heralded as "newly reinvigored naturality" (1587), but this heralding has a lot of detractors—most of them livestock farmers in the region. The problem, from Buller's perspective, is that both proponents and opponents of the wolves construct them as "purified 'outsiders': entirely emblematic of an externalized, and thus essentially dualistic, conception of nature and its relation to human society" (1591). This construction leads to a false separation of humans from nature and also tries to falsely bound the natural away from the human by trying to contain the wolves to specific areas—in essence, as Lulka argues in the case of bison, control their mobility and fix them in space.

Much of the politics around geographies of human-wildlife relations have to do with who gets to use, protect, or otherwise control wildlife for preservation, conservation, or use as a resource. If we take the case of sea turtle conservation we can see how "it is articulated and executed at different sociopolitical and geographic scales, and the consequences for local rights of access to resources" (L. Campbell 2007, 313). In focusing their case study on sea turtles in the Caribbean, Campbell and Godfrey (2010) also highlight the difficulty in managing such a mobile population of animals, which some groups want protected and other groups want to consume as a resource. As a result, "promoting conservation action at a particular scale is not simply a matter of biological or ecological necessity, but serves the political interests of

particular groups" (L. Campbell 2007, 313). "The question of the appropriate scale at which sea turtle populations should be conceptualized is central
to debates about sea turtle conservation more generally" (Campbell and
Godfrey 2010, 898). What has happened with sea turtle conservation is that
the management position opposes consumptive use at the international scale,
and this position is translated down to the national and then local scales no
matter which communities might want or need to consumptively use sea
turtles. With the case of the hawksbill sea turtles in the Caribbean, which are
harvested by Cubans for sale to Japan for use in fashion and art, "one of the
challenges for understanding sea turtle ecology has been the difficulty of linking individual turtles found at different life stages and using distinct habitats
to larger populations" (899). In this case, opponents of the sea turtle harvest
by Cubans argued the Cubans did not have the right to use these animals since
doing so impinged on the rights of other countries in the region to utilize them
as a resource or conserve them. But the Cubans argued that the turtles were
"Cuban," and therefore, they were within their rights to utilize species found
in their political territory. Using genetic testing to study the movement of the
turtles begins to reveal that the turtles are regional and that both overlap and
separation occur in the populations. The importance of a geographic perspective here is that the scale of genetics is being used to influence the broad-scale
geopolitics of nature conservation, and we see how the construction of the
"the natural" serves as a "socio-spatial" disciplining mechanism.

The politics of hunting can be quite contentious, pitting those who find
killing animals wrong and brutal against those who see hunting as a survival
tool or a way to get closer to the animals. In an analysis of the debate over
fox hunting in the United Kingdom in 1997 brought about by the Foster Bill,
a proposal to ban fox hunting with dogs, Michael Woods (2000) shows how
the representation of the fox involved three conflicting ideas of the fox as
sporting foe, pest, and victim. In the first representation the fox was seen to be
an equal and cunning contestant almost wanting to participate in its own hunt.
Fox hunting normally takes place with a pack of dogs who are let loose and
roam until they catch the scent of a fox, and then they hunt it down until the
humans ride up on their horses at the end to determine what should happen.
At the same time, hunters depicted the fox not as an equal but as a vicious
vermin that needed to be controlled. With this view the hunters were actually
doing a public service by killing foxes so they wouldn't raid human territory.
Finally, anti-fox-hunting advocates represented the fox as a victim. They argued that fox hunting is a violation of nature and uncivil in today's time. They
drew upon scientific studies that showed the stress of hunting on foxes and
represented it as inhumane. Woods emphasizes that representation is not just
a re-presentation but a translation of the fox by humans, and even as foxes

could not participate in the political process, they asserted themselves into national and local politics, thereby forcing multiple stakeholders to consider who these foxes were.

With respect to theoretical frameworks, Jennifer Wolch has done much to bring wildlife into the folds of the urban and point outs that "although urbanization has further distanced people from nature, this very distanciation has in part fueled a resurgent biophilia" (Wolch, West, and Gaines 1995, 736). In developing the concept of transspecies urban theory, Wolch, West, and Gaines argue such a theory is needed to understand four issues: how urbanization impacts wildlife, how and why residents react to wildlife the way they do, how city building practices shape urban ecologies, and how urban planning/policy can better incorporate wild animals. They point out that both extreme social and extreme spatial fragmentation cause problems. Not only is the habitat for wildlife in urban areas incredibly fragmented posing myriad dangers to animals from crossing busy streets to locating sufficient food sources, but from an anthropocentric perspective urban wildlife usually fall into the social categories of either pests or pets. Jennifer Wolch also calls for the formation of a "zoöpolis," because "in mainstream theory, urbanization transforms 'empty' land through a process called 'development' to produce 'improved land,' whose developers are exhorted (at least in neoclassical theory) to dedicate it to the 'highest and best use'" (1995, 119). Therefore, "the recovery of animal subjectivity implies an ethical and political obligation to redefine the urban problematic and to consider strategies for urban praxis from the standpoints of animals" (122). The challenge for a transspecies urban theory is to attempt to remedy this fragmentation in both ways—through education, zoning, land acquisitions, environmental impact statements, and wildlife fees. In essence, Wolch, West, and Gaines are arguing that societal commitment to biodiversity and wildlife conservation/protection cannot stop at the city gates.

Clare Palmer (2003) builds on Wolch and argues that the metaphor of colonization, because of its inherent processes of dispossession, negotiation, and resistance, is useful in thinking through urban human-animal relations. For example, the concept of development and its manifestation on the landscape for animals parallels the experiences of colonized peoples who were made invisible by colonizers who saw the land they wanted as empty. "In the cases of both colonized human and animals, invisibility and distancing allow land to be occupied [or developed] despite the previous existence of inhabitants" (50). Does this mean then that all disposed animals are powerless? Palmer takes pains to point out that animals do have different types of power in the form of resistance: they can leave the area, stay and survive alongside humans, or stay and utilize the human environment for survival. Palmer concludes by saying that examining the full gamut of power relations between humans and wildlife in urban areas is essential to help create new transspecies practices for a zoöpolis.

Jamie Lorimer (2008) traces the trials of urban brownfield and living roof conservationists as they advocate for these spaces to be used for wildlife in the city. In his case study of brownfields—former industrial areas that have been abandoned—he finds three main problems. The first is persuading people to see that wildlife lives in these places at all because most people are so conditioned to thinking of them as lost spaces. The second is that brownfield sites have also been seen as eyesores, and any wildlife surviving in them become out of place. People feel that in order to have wildlife we need to "greenwash" these areas and turn them into parklike spaces that are heavily managed rather than allow them to exist on their own terms. Finally, he argues that the wildlife that inhabit these already disliked places do not carry enough charisma to attract advocates. He urges us to reconsider these urban spaces as more fluid where there is a "focus on difference, rather than diversity" (2055).

In another case study of the urban human-wildlife interface, Gullo, Lassiter, and Wolch focus on cougars in California to examine how human ideas about the cougar have been "shaped by patterns of urbanization, by scientific and political debate, and by changing media coverage" (1998, 139). One of their key points is that "in so-called modern societies, the social construction of animals goes largely unmediated by concrete experience, lending the social imaginary even greater constitutive force" (140). After all, how many of us have really ever had a personal interaction with a cougar? Most of what we know and subsequently think comes from other people's information. In Orange County, California, urban sprawl has increased dramatically over the past couple of decades and human communities are pushing further and further into what was traditionally cougar habitat—leading to more cougar-human encounters. Some of the encounters are benign and even enjoyable for humans who like to watch wildlife while others find living in such close proximity to cougars terrifying mainly because of fears about safety for children and domestic pets. As the authors move through the different ways stakeholders frame these fears and desires, they conclude that humans really need to be educated about best practices to live with cougars in order to provide firsthand knowledge for humans to draw from rather than relying solely on third-party information. Again, we see this notion of wildlife needing to be in its proper place: wildlife is acceptable as long as it stays within the boundaries set for it by humans regardless of whether or not humans themselves might be out of place in wildlife habitat. Cougars themselves have become political agents in their own way—no, they do not vote, protest, or show up at town meetings—but they have changed their behaviors, moving at different times of day, moving back and forth between developed and undeveloped areas, and even standing their ground.

The politics of human-wildlife conflict (HWC), our third area of human-wildlife relations, encompasses a wide variety of geographic locations and mainly involves either property damage/loss from wild animals or direct threats to human safety (Treves et al. 2006). Indeed, "the sociopolitical setting is as influential as the biophysical one" (384). In a key paper on animal geographies and human-wildlife conflict, JoAnn McGregor (2005) uses the case of the listing of the crocodile as endangered despite localized resistance to show how considering both the attitudes and circumstances of local peoples who bear the physical and economic costs of living with dangerous animals is key to establishing successful human-wildlife policies. In turning one of the most feared animals in all of Africa into one worthy of international concern and attention, "the new conservationist image of the Nile crocodile was promoted in the context of globalized networks of commercial interest in crocodile skins that had initially encouraged that animal's decimation but were subsequently implicated in its recovery" (306). The reconfiguration took place by an international (and northern) community that did not attend to local belief systems or realize the difficulties of living with these animals. Instead, these local concerns were simply dismissed. "Post-colonial relations of power, and the precarious nature of local livelihoods pose a profound challenge to the idea of 'bringing the animals back in'—both imaginatively and practically" (306).

Jun-Han Yeo and Harvey Neo (2010) document one case of humans in conflict with long-tailed macaques in and near the Bukit Timah Nature Reserve in Singapore. This nature reserve is a borderland community where humans and wild animals share spaces. The macaques have normally been culled to try and control the numbers, but people still complain that they get into trash, steal food and household items, make too much noise, and generally are a nuisance by not staying "in their place." The National Parks Board has to constantly negotiate between supporting biodiversity and complaining residents. The residents, who hold "dwelt" perspectives in the sense that they have daily encounters with the macaques are much more vocal in constructing them as out of place because they "disrupt the context of home as a safe, secure, autonomous territory" (690). These residents argue the macaques should be placed elsewhere, but this elsewhere is becoming more and more problematic as the area right around the reserve becomes more and more urbanized. Contrary to residents' dwelt experiences, those professionals that are asked to mediate—either through education, by directly removing a problem animal, or by responding to complaints—by mobilizing a discourse that constructs the macaques as "in place" and vital to the native ecosystem.

A case study of human-wildlife conflict in a rural area comes from work by Monica Ogra (2008) in northern India. She argues that both the visible and hidden costs of human-wildlife conflict need to be addressed in order to implement better policies for both humans and animals. In the village of Bhalalogpur, wildlife damage crops, prey on livestock, damage property, and also attack humans, not only making life harder for low-income rural communities but also undermining support for conservation and in many cases leading to "retaliation killings" upon wildlife. Ogra points out that most work on human-wildlife conflict focuses on establishing methods of avoidance or compensation, exposing various political and cultural challenges of attitudes toward wildlife, analyzing basic causes and effects of human-wildlife encounters, and understanding HWC as a problem of the poor. She says that geographers need to go deeper to understand the hidden costs and gendered components. She finds in her case study that damage by wildlife causes a high level of emotional trauma to all individuals, an increased workload when men have to fix damaged buildings or women have to use different routes to gather wood and water to avoid an aggressive animal, a loss of labor when someone is harmed, which can have detrimental impact on already struggling families, and a decrease in nutrition, especially for women, when crops are destroyed or consumed, leaving not enough food to go around. While focusing almost exclusively on the hidden costs for the human community rather than the costs for wildlife, Ogra highlights that living in close quarters with wildlife means ensuring that as much as possible is taken into consideration in each place to address the needs of the people in order to ensure safety and survival for both humans and wildlife.

Which land gets used for wildlife protection is also a political struggle. David Havlick (2011) presents a case study of one site of military-to-wildlife conversion in Indiana and reflects on the implications of casting military practices and environmental conservation as compatible activities. The land in question had been used since WWII as a bombing range, and unexploded live ordinance is still strewn across the landscape. Instead of paying to keep it as military space, the military decided to turn it over for conservation land. This piece of land has become an important addition to the National Wildlife Refuge system, but it is also some of the most contaminated land there is (military landscapes, that is). Havlick argues this land use switch from military to wildlife/natural environment poses a challenge to the idea of wildlife refuges as pristine natural spaces and opens up new notions of hybrid geographies by blurring the distinctions between categories of land use and notions of the pristine.

Given the myriad human-wildlife geography issues, how do geographers think things can be done differently? The politics of engagement is our fourth area of emphasis for this section. Steve Hinchliffe and colleagues (2005) argue that this can best be accomplished through experiential field practices that can help develop new understandings of animal agency and in turn lead to new political practices that can absorb the complexity of human-animal entanglements. They call this "cosmopolitics." "Ecologizing politics is not about producing better or more accurate representations, as if we can take preexisting identities and bring them into the conversation. Rather, it is about changing engagements" (651). For their project they engaged with water voles, small rodents, by learning how to read their signs—tracks, scat, dens— all without coming into direct contact with these largely nocturnal, secretive animals. In the practice of changing engagements, in learning how to "see" or understand water voles in their own places, perhaps humans can come to see how cohabitation is possible and to actually desire it because "cultures and societies are shaped by more than human geographies" (644). Peter Yaukey (2010) shows that getting the public to engage with other species is a parallel way of practicing cosmopolitics. He sees a strategy to both further biological and cultural knowledge of other species in enlisting the help of amateur naturalists to collect data on a large scale. He shows that this approach has found success with birds through such events as the Audubon Christmas bird count—the largest bird-count effort on the globe—and Project FeederWatch by the Cornell Lab for Ornithology. These practices succeed because they elicit a response by the general public toward nonhuman species allowing humans to both participate in the process of knowledge generation for humans and gain experience in identifying and considering the lives of others. Either way, cosmopolitics involves an active doing on the part of humans to engage with nonhuman others.

A related example of active doing to engage with nonhuman others comes from Suzanne Michel (1998), who uses the concept of ethics of care from feminist theories to talk about how alternative places for human-wildlife encounters can shape new ways of being with nonhumans. An ethics of care can be understood as an ethics that recognizes right ways of being in the world, including caring for others—something traditionally done by women in many societies around the world. This idea can transfer to wildlife because it fosters "nondualistic thinking, which allows local communities and individuals to become experientially and emotionally connected with the plight of disappearing wildlife" (163). Caring about wildlife differs from scientific or political management of wildlife because in these instances decisions are supposed to be made based on rational and scientific data, which "denies human and nonhuman agency, [and] the importance of individuals in the creation and transformation of our nature-society relations and landscapes" (169). She argues for the political and ethical importance of spaces like wildlife

rehabilitation centers and environmental education in public spaces. Seeing
these two locations as borderlands—areas in which boundaries are blurred
and places where "conventional approaches are questioned, stereotypes dissolve, and new understandings emerge" (162)—she believes that they can
also be places where an ethics of care for nonhumans can be tested out and
developed. Wildlife rehabilitation and environmental education offer a place
where notions of care can be reconceptualized as political action and kinship
with other species. In the case of golden eagles she finds that people are much
more receptive to considering how to live with and care for these large raptors
once they get a chance to "know" them.

Returning to Sri Lanka's human-elephant relationships, Jamie Lorimer
(2010) calls for a convivial biogeography that includes three points that
could change the politics of human-wildlife interactions. First, he calls for
recognition of nonhuman difference and awareness of the ways in which
nonhuman companionship is forged. Second, a deeper attention to interspecies conviviality—the ways in which species interact—will be necessary for
envisioning new relations. Third, a "cosmopolitan environmentalism" echoes
Wolch's transspecies urban theory in calling for finding common political,
policy, and scientific intersections that provide space for a more-than-human
world. He does this by documenting the history of human-elephant relations
in Sri Lanka, which have, in fact, been relations that have coevolved over
hundreds of years. Even something as intimate as tuberculosis—which can
pass from humans to elephants—has shaped historical relations. The arrival
of the British disrupted local relations, and today human-elephant relations
are constituted by different sets of interests—those of mainly British tourists
who desire digital trophies and demand to see "wild" elephants, local farmers
who live with the very real threat of elephant raids on their fields, local ma-
houts who live and work with elephants on a daily basis getting to know them
quite intimately, the scientific community who wishes to study and conserve
elephants, and the general public both in Sri Lanka and abroad who may or
may not worry about them.

Jamie Lorimer also maps the concept of nonhuman charisma to "forge a
more-than-human understanding of agency and to consider the ethical implications of this realignment" because "in our contemporary world of avian
flu, genetic modification, and climate change, nonhuman agency is both a
common sense observation and a tautology" (2007, 912). For Lorimer, nonhuman charisma can best be defined as the "distinguishing properties of a
non-human entity or process that determine its perception by humans and its
subsequent evaluation" (915). His fundamental argument is that humans—
who inhabit limited human bodies, have access to limited technologies, and
inhabit different cultural contexts—necessarily shape nonhuman charisma in
terms of its perception by humans. Indeed, "we can see how the physiologi-

cal and phenomenological configuration of the human body puts in place a range of filtering mechanisms that disproportionately endow certain species with ecological charisma" (916). He highlights three types of charisma—the ecological, the aesthetic, and the corporeal. In the first case, the more an animal's ecological rhythms (diet, movement, activities) resemble humans, the more we will endow that animal as having charisma—hence our construction of wolves, not centipedes, as having charisma. Secondly, aesthetic charisma comes from appearance and behavior that trigger strong emotional responses in humans. We respond to chimpanzees and gorillas because they look so much like us—their eyes, their hands, their movements are all accessible to our sensory experiences as humans—whereas the appearance and behaviors of an animal such as an earthworm or a sea urchin are so far removed from triggering an emotional response that these species become charismatic outsiders. Consider the WWF's panda logo—pandas with their soft fur, rotund shapes, small ears, and comic markings trigger a response from humans—but had the WWF chosen a sea urchin as their logo, would the marine animal have had the same draw? Lorimer would say no. Finally, what Lorimer refers to as corporeal charisma has to do with the affections and emotions engendered by different species and their interactions with humans over time—what he terms *interspecies epiphanies*, during which humans have learned to "see" other species and have some type of relation with them. Wolves, elephants, tigers, deer, cats, dogs, cows, and pigs would all be part of this equation. Because of our relationships with them—both positive and negative—we see them differently (and at all). Therefore, our ability to experience a more-than-human world and the ethical questions that arise rely on our own physicality as humans. He argues that this recognition must be brought to the fore as we seek ethical relations with a more-than-human world.

In this chapter we have explored the extensive geographies of human-wildlife relations. We should now have a good understanding of the ways in which we construct wildlife, the ways in which we consume them literally as food, medicine, or accessories or experientially as tourists, and the complex ethical/political conflicts that shape where and how wildlife can live.

DISCUSSION QUESTIONS

1. Is "camera hunting" the same as "real" hunting? How are they identity-driven practices?
2. What human-wildlife conflicts exist in your particular place? How is the conflict framed? By whom? How can you distinguish the links among place, culture, politics, economics, and ethics?
3. Is "wild" a useful concept?

KEYWORDS/CONCEPTS

biodiversity
biophilia
camera hunting
CITES
cosmopolitics
ecotourism
ethics of care
human-wildlife conflict

ornithophilia
species charisma
transspecies urban theory
wild
wildlife
zoogeomorphology
zoöpolis

PRACTICING ANIMAL GEOGRAPHY

1. Examine the images of wildlife in your area using magazines, advertisements, and visits to local wildlife parks, centers, and sanctuaries. What
   constructions of the wild do you find? Why?
2. Conduct a weeklong field study of the wildlife around your home and in
   your daily life. Keep a notebook, and note day, time, and location, species,
   all behaviors, and your reflections.

RESOURCES

CITES: http://www.cites.org
Convention on Biological Diversity: http://www.cbd.int
Global Ecology and Biogeography: http://www.wiley.com/bw/journal.asp?ref=1466
   -822X
Karl Ammann (documentary filmmaker with films about wildlife trade and bush-
   meat): http://karlammann.com
Keiko: The Untold Story: http://www.keikotheuntoldstory.com
Human Dimensions of Wildlife: http://www.tandf.co.uk/journals/titles/10871209.asp
International Biogeography Society: http://www.biogeography.org
International Union for the Conservation of Nature: http://www.iucn.org
John Downer Productions (films about wildlife focusing on subjectivity): http://jdp
   .co.uk
Journal of Biogeography: http://www.blackwellpublishing.com/journal.asp?ref
   =0305-0270
Milking the Rhino (documentary about conservation and living with wildlife in Af-
   rica): http://milkingtherhino.org/film.php
Millennium Ecosystem Assessment: http://www.maweb.org/en/Index.aspx
Planet Earth (documentary about life on earth): http://dsc.discovery.com/tv/planet
   -earth

TRAFFIC: http://www.traffic.org
US Fish and Wildlife Service: http://www.fws.gov
World Wildlife Fund: http://www.worldwildlife.org

REFERENCES

Anderson, Kay. 2000. "'The Beast Within': Race, Humanity, and Animality." *Environment and Planning D: Society and Space* 18:301–320.

———. 2003. "White Nature: Sydney's Royal Agricultural Show in Post-humanist Perspective." *Transactions of the Institute of British Geographers* 28:422–441.

Associated Press. 2010. "Putin Praises DiCaprio as Tiger Pledge Signed." Accessed June 1, 2011. http://www.cbsnews.com/stories/2010/11/23/tech/main7082817.shtml.

Baer, Leonard D., and David R. Butler. 2000. "Space-Time Modeling of Grizzly Bears." *The Geographical Review* 90 (2): 206–221.

Bear, Christopher, and Sally Eden. 2008. "Making Space for Fish: The Regional, Network and Fluid Spaces of Fisheries Certification." *Social and Cultural Geography* 9 (5): 487–504.

———. 2011. "Thinking like a Fish? Engaging with Nonhuman Difference through Recreational Angling." *Environment and Planning D: Society and Space* 29:336–352.

Besio, Kathryn, Lynda Johnston, and Robyn Longhurst. 2008. "Sexy Beasts and Devoted Mums: Narrating Nature through Dolphin Tourism." *Environment and Planning A* 40:1219–1234.

Bonta, Mark. 2010. "Ornithophilia: Thoughts on Geography in Birding." *The Geographical Review* 100 (2): 139–151.

Brownlow, Alec. 2000. "A Wolf in the Garden: Ideology and Change in the Adirondack Landscape." In *Animal Spaces, Beastly Places: New Geographies of Human-Animal Relations*, edited by Chris Philo and Chris Wilbert, 141–158. New York: Routledge.

Buller, Henry. 2004. "Where the Wild Things Are: The Evolving Iconography of Rural Fauna." *Journal of Rural Studies* 20:131–141.

———. 2008. "Safe from the Wolf: Biosecurity, Biodiversity, and Competing Philosophies of Nature." *Environment and Planning A* 40:1583–1597.

Cammack, Paul J., Ian Convery, and Heather Prince. 2011. "Gardens and Birdwatching: Recreation, Environmental Management and Human-Nature Interaction in an Everyday Location." *Area* 43 (3): 314–319.

Campbell, Lisa M. 2007. "Local Conservation Practice and Global Discourse: A Political Ecology of Sea Turtle Conservation." *Annals of the Association of American Geographers* 97 (2): 313–334.

Campbell, Lisa M., and Matthew H. Godfrey. 2010. "Geo-political Genetics: Claiming the Commons through Species Mapping." *Geoforum* 41 (6): 897–907.

Campbell, Michael O'Neal. 2007. "An Animal Geography of Avian Ecology in Glasgow." *Applied Geography* 27:78–88.

Campbell, Michael O'Neal, and Maria Elena Torres Alvarado. 2011. "Public Percep-tions of Jaguars *Pantera onca*, Pumas *Puma concolor*, and Coyotes *Canis latrans* in El Salvador." *Area* 43 (3): 250–256.

Carter, Jenny. 1997. "Nest-Site Selection and Breeding Success of Wedge-Tailed Shearwaters *Puffinus pacificus* at Heron Island." *Australian Geographical Studies* 35 (2): 153–167.

Castree, Noel. 1997. "Nature, Economy and the Cultural Politics of Theory: The 'War against the Seals' in the Bering Sea, 1870–1911." *Geoforum* 28 (1): 1–20.

Cloke, Paul, and Harvey C. Perkins. 2005. "Cetacean Performance and Tourism in Kaikoura, New Zealand." *Environment and Planning D: Society and Space* 23:903–924.

Connell, John. 2009. "Birdwatching, Twitching and Tourism: Towards an Australian Perspective." *Australian Geographer* 40 (2): 203–217.

Convention on International Trade in Endangered Species of Wild Fauna and Flora. 2011. "The CITES Export Quotas." Accessed April 14. http://www.cites.org/eng/resources/quotas/index.shtml.

Crifasi, Robert R. 2007. "A Subspecies No More? A Mouse, Its Unstable Taxonomy, and Western Riparian Resource Conflict." *Cultural Geographies* 14:511–535.

Davies, Gail. 1999. "Exploiting the Archive: And the Animals Came in Two by Two, 16mm, CD-ROM and BetaSp." *Area* 31 (1): 49–58.

———. 2000. "Science, Observation and Entertainment: Competing Visions of Post-war British Natural History Television, 1946–1967." *Ecumene* 7 (4): 432–460.

Duffus, David A. 1996. "The Recreational Use of Grey Whales in Southern Clayo-quot Sound, Canada." *Applied Geography* 16 (3): 179–190.

Duffy, Rosaleen, and Lorraine Moore. 2010. "Neoliberalising Nature? Elephant-Back Tourism in Thailand and Botswana." *Antipode* 42 (3): 742–766.

Elder, Glen, Jennifer Wolch, and Jody Emel. 1998. "Le Pratique Sauvage: Race, Place, and the Human-Animal Divide." In *Animal Geographies*, edited by Jennifer Wolch and Jody Emel, 72–90. New York: Verso.

Emel, Jody. 1998. "Are You Man Enough, Big and Bad Enough? Wolf Eradication in the US." In *Animal Geographies*, edited by Jennifer Wolch and Jody Emel, 91–116. New York: Verso.

Gade, Daniel W. 2006. "Hyenas and Humans in the Horn of Africa." *The Geographi-cal Review* 96 (4): 609–632.

———. 2010. "Shifting Synanthropy of the Crow in Eastern North America." *The Geographical Review* 100 (2): 152–175.

Gillespie, Thomas W. 2001. "Remote Sensing of Animals." *Progress in Physical Geography* 25 (3): 355–362.

Gullo, Andrea, Unna Lassiter, and Jennifer Wolch. 1998. "The Cougar's Tale." In *Animal Geographies*, edited by Jennifer Wolch and Jody Emel, 139–161. New York: Verso.

Havlick, David G. 2011. "Disarming Nature: Converting Military Lands to Wildlife Refuges." *The Geographical Review* 101 (2): 183–200.

Herrman, Bernd, and William I. Woods. 2010. "Neither Biblical Plague nor Pristine Myth: A Lesson from Central European Sparrows." *The Geographical Review* 100 (2): 176–186.

Hinchliffe, Steve. 2007. *Geographies of Nature: Societies, Environments, Ecologies.* London: Sage.

Hinchliffe, Steve, Matthew B. Kearnes, Monica Degen, and Sarah Whatmore. 2005. "Urban Wild Things: A Cosmopolitical Experiment." *Environment and Planning D: Society and Space* 23:643–658.

International Union for the Conservation of Nature. 2011. "Red List Summary Statistics." Accessed October 3. http://www.iucnredlist.org/about/summary-statistics.

Laudati, Ann. 2010. "Ecotourism: The Modern Predator? Implications of Gorilla Tourism on Local Livelihoods in Bwindi Impenetrable National Park, Uganda." *Environment and Planning D: Society and Space* 28:726–743.

Lorimer, Jamie. 2006. "What about the Nematodes? Taxonomic Partialities in the Scope of UK Biodiversity Conservation." *Social and Cultural Geography* 7 (4): 539–558.

———. 2007. "Nonhuman Charisma." *Environment and Planning D: Society and Space* 25:911–932.

———. 2008. "Living Roofs and Brownfield Wildlife: Towards a Fluid Biogeography of UK Nature Conservation." *Environment and Planning A* 40:2042–2060.

———. 2010. "Elephants as Companion Species: The Lively Biogeographies of Asian Elephant Conservation in Sri Lanka." *Transactions of the Institute of British Geographers* 35:491–506.

Lorimer, Jamie, and Sarah Whatmore. 2009. "After the 'King of Beasts': Samuel Baker and the Embodied Historical Geographies of Elephant Hunting in Mid-Nineteenth-Century Ceylon." *Journal of Historical Geography* 35:668–689.

Lulka, David. 2004. "Stabilizing the Herd: Fixing the Identity of Nonhumans." *Environment and Planning D: Society and Space* 22:439–463.

———. 2008. "The Intimate Hybridity of Roadkill: A Beckettian View of Dismay and Persistance." *Emotion, Space and Society* 1:38–47.

Matless, David. 2000. "Versions of Animal-Human: Broadland, c. 1945–1970." In *Animal Spaces, Beastly Places: New Geographies of Human-Animal Relations*, edited by Chris Philo and Chris Wilbert, 115–140. New York: Routledge.

Matless, David, Paul Merchant, and Charles Watkins. 2005. "Animal Landscapes: Otters and Wildfowl in England 1945–1970." *Transactions of the Institute of British Geographers* 30:191–205.

McGregor, JoAnn. 2005. "Crocodile Crimes: People versus Wildlife and the Politics of Postcolonial Conservation on Lake Kariba, Zimbabwe." *Geoforum* 36 (3): 353–369.

Michel, Suzanne M. 1998. "Golden Eagles and the Environmental Politics of Care." In *Animal Geographies*, edited by Jennifer Wolch and Jody Emel, 162–187. New York: Verso.

Millennium Ecosystem Assessment. 2005. *Ecosystems and Human Well-Being: Biodiversity Synthesis.* Washington, DC: World Resources Institute.

Mitchell, Martin D., Richard O. Kimmel, and Jennifer Snyders. 2011. "Reintroduction and Range Expansion of Eastern Wild Turkeys in Minnesota." *The Geographical Review* 101 (2): 269–284.

Naughton-Treves, Lisa. 2002. "Wild Animals in the Garden: Conserving Wildlife in Amazonian Agroecosystems." *Annals of the Association of American Geographers* 92 (3): 488–506.

Neves, Katja. 2010. "Cashing In on Cetourism: A Critical Ecological Engagement with Dominant E-NGO Discourses on Whaling, Cetacean Conservation, and Whale Watching." *Antipode* 42 (3): 719–741.

Norton, Andrew. 1996. "Experiencing Nature: The Reproduction of Environmental Discourse through Safari Tourism in East Africa." *Geoforum* 27 (3): 355–373.

Ogra, Monica. 2008. "Human-Wildlife Conflict and Gender in Protected Area Borderlands: A Case Study of Costs, Perceptions, and Vulnerabilities from Uttarakhand (Uttaranchal), India." *Geoforum* 39 (3): 1408–1422.

Palmer, Clare. 2003. "Colonization, Urbanization, and Animals." *Philosophy and Geography* 6 (1): 47–58.

Power, Emma R. 2009. "Border-Processes and Homemaking: Encounters with Possums in Suburban Australian Homes." *Cultural Geographies* 16:29–54.

Proctor, James D. 1998. "The Spotted Owl and the Contested Moral Landscape of the Pacific Northwest." In *Animal Geographies*, edited by Jennifer Wolch and Jody Emel, 191–217. New York: Verso.

Robbins, Paul, John Hintz, and Sarah A. Moore. 2010. *Environment and Society: A Critical Introduction*. Malden, MA: Wiley-Blackwell.

Ryan, James R. 2000. "'Hunting with the Camera': Photography, Wildlife and Colonialism in Africa." In *Animal Spaces, Beastly Places: New Geographies of Human-Animal Relations*, edited by Chris Philo and Chris Wilbert, 203–221. New York: Routledge.

Suchet, Sandie. 2002. "'Totally Wild'? Colonising Discourses, Indigenous Knowledges and Managing Wildlife." *Australian Geographer* 33 (2): 141–157.

Treves, Adrian, Robert B. Wallace, Lisa Naughton-Treves, and Andrea Morales. 2006. "Co-managing Human-Wildlife Conflicts: A Review." *Human Dimensions of Wildlife* 11:383–396.

US Fish and Wildlife Service. 2006. *National Survey of Fishing, Hunting, and Wildlife-Associated Recreation*. Washington, DC: US Department of the Interior.

Vaccaro, Ismael, and Oriol Beltran. 2009. "Livestock versus 'Wild Beasts': Contradictions in the Natural Patrimonialization of the Pyrenees." *The Geographical Review* 99 (4): 499–516.

Waley, Paul. 2000. "What's a River without Fish? Symbol, Space and Ecosystem in the Waterways of Japan." In *Animal Spaces, Beastly Places: New Geographies of Human-Animal Relations*, edited by Chris Philo and Chris Wilbert, 159–181. New York: Routledge.

Whatmore, Sarah, and Lorraine Thorne. 1998. "Wild(er)ness: Reconfiguring the Geographies of Wildlife." *Transactions of the Institute of British Geographers* 23:435–454.

———. 2000. "Elephants on the Move: Spatial Formations of Wildlife Exchange." *Environment and Planning D: Society and Space* 18:185–203.

Wilson, Edward O. 1984. *Biophilia.* Cambridge: Harvard University Press.

Wilson, Robert M. 2009. "Birds on the Home Front: Wildlife Conservation in the Western United States during World War II." In *War and the Environment: Military Destruction in the Modern Age*, edited by Charles E. Closmann, 132–149. College Station: Texas A&M Press.

Withgott, Jay, and Scott Brennan. 2008. *Environment: The Science behind the Stories.* San Francisco: Pearson Benjamin Cummings.

Wolch, Jennifer. 1995. "Zoopolis." *Capitalism, Nature, Socialism* 7 (2): 21–47.

Wolch, Jennifer, Kathleen West, and Thomas Gaines. 1995. "Transspecies Urban Theory." *Environment and Planning D: Society and Space* 13 (6): 735–760.

Wolch, Jennifer, and Jin Zhang. 2005. "Siren Songs: Gendered Discourses of Concern for Sea Creatures." In *A Companion to Feminist Geography*, edited by Lise Nelson and Joni Seager, 458–485. Oxford: Blackwell.

Woods, Michael. 2000. "Fantastic Mr. Fox? Representing Animals in the Hunting Debate." In *Animal Spaces, Beastly Places: New Geographies of Human-Animal Relations*, edited by Chris Philo and Chris Wilbert, 182–202. New York: Routledge.

Yaukey, Peter H. 2010. "Citizen Science and Bird-Distribution Data: An Opportunity for Geographic Research." *The Geographical Review* 100 (2): 263–273.

Yeo, Jun-Han, and Harvey Neo. 2010. "Monkey Business: Human-Animal Conflicts in Urban Singapore." *Social and Cultural Geography* 11 (7): 681–699.

Chapter Seven

Conclusion: The Place of Geography in Human-Animal Studies

This book has been a whirlwind introduction to the intersection of geography and human-animal studies known as animal geography. After a brief summary of the book, this last chapter will step back in order to provide some final thoughts and reflections on future directions for animal geography. A final "practicing animal geography" exercise will allow you to synthesize all that we have covered.

WHERE WE HAVE BEEN

Chapter 1 identified four key social changes that have contributed to the rise of the third wave of animal geography over the past fifteen years: our deepening understanding of how humans are impacting the natural world, the rise of animal-related social movements, the theoretical shift to a postmodern/posthuman framework that is learning to see other-than-human beings as actors in the world, and finally our increasingly public love of nonhumans. The result of these shifts for academia has been the rise of human-animal studies—an interdisciplinary and multidisciplinary effort to unpack and examine the myriad relations humans have had with nonhuman animals.

This book has aimed to explore animal geography to gain an understanding of how geography can contribute to the human-animal studies project even as animal geography enriches its own home discipline. It has synthesized for the first time the existing body of work that is animal geography by framing it through major human geographical analytic categories (economic, ethical, historical, and political geographies along with the cultural landscape) and concepts (landscape, place, power, scale, space). The book began with an introduction to both geography and human-animal studies in order to lay the groundwork for the need for a specific focus on animals within geography and a need for incorporating geography into human-animal studies.

The second chapter provided an overview of the history of the three waves of animal geography that have expanded and built on one another. The earliest animal geography began with cataloging species and evolutionary adaptations. This part of animal geography continues today within the subdiscipline of biogeography and is a key methodological and data-driven research area that is helping to directly understand animal behaviors and their environmental links. As the second wave of animal geography appeared, the focus shifted away from cataloging species to understanding how domesticated livestock shaped, and are shaped by, human cultures and the landscape. This expansion of animal geography also carries through to the present and is most often seen in cultural and political ecology research. The third wave of expansion for animal geography came in the mid-1990s as a reaction to those four key social changes listed above. What could be studied under the discipline of animal geography expanded dramatically to encompass all forms of human-animal relationships, not just those involved in livestock or wildlife mapping. In fact, the third wave of animal geography has made it clear that not only is the natural world populated by sentient, individual beings but our human societies are awash in animals—whether through their parts, which we wear and eat, their images, which we consume as media or toys, their living selves, which we share our intimate spaces with, or their labor, which we make work for us.

The next four chapters each focused on one of the major umbrella human-animal categories: pets and culture, working animals, farmed animals, and wild animals. In surveying these categories we focused on showing how animal geographers have illuminated particular relations in particular places. We found that, across the board, each umbrella category has its own history, economy, impact on the cultural landscape, and ethical/political issues. In fact, the one constant sustained throughout the different categorical lenses is the phenomenal extent to which nonhumans are intertwined with human lives. Whether at the intimate scale of the home and body or at the global scale with the impact on climate change from industrially raised livestock, no corner of human society is without nonhuman influence.

The role of place is, perhaps, the most fundamental idea that emerges from the body of work that is animal geography. We have built a conception of place that includes not only the physical realities (e.g., farm, zoo, home) but also the conceptual locations of animals (e.g., pet, pest, food). When it comes to animals, we cannot understand one without the other. For example, industrial farming and the places that are CAFOs in the landscape only make sense because the animals that exist in them are conceptually placed as food animals. In terms of animal geography's contribution to the larger human-animal studies community, the full spectrum of place as a conceptual tool is the most important. An animal geography perspective reminds us that we cannot simply talk about nonhuman animals, but we must instead go out and meet them in their locations and as individuals and breeds or species. To understand the human-animal interspecies power geometry we must first map the place relations that shape practices in the first place. The contributions of animal geography to the larger discipline of geography are also important. While we have shown how animals have been visible over the history of geography, we have also shown how they haven't been "seen" as sentient, individual beings. Human geography has barely acknowledged the existence of animals, and biogeography has barely recognized the human-animal relationship and the individual subjective experiences of the animals themselves. This book provides evidence that nonhumans play a much larger role in human identity formation, landscape practices, and political conflicts than has heretofore been recognized.

THE FUTURE OF ANIMAL GEOGRAPHY

This book, while a synthesis of animal geography, is itself a snapshot of a certain place and time. It is not meant to provide the defining structure of animal geography as it moves forward, but the foundation. Where is animal geography going? The future is promising for both an expansion of the topics covered but also the research methodologies. As we have already seen, many more topics are left to study than what animal geographers have done, and indeed probably many have not even been mentioned in the book.

We have seen how the focus, thus far, has been mainly on wildlife and then livestock animals. One of the first ways animal geography can branch out is to challenge itself to see and then study the full spectrum of human-animal relations. It has often admonished others in geography for keeping animals hidden, but now it must engage with itself more directly. From entertainment animals to human identity formation around animals, enough geography can be found to take the subfield well into the future. Part of this research trajectory will benefit from a focus on the geographies of cultural attitudes toward animals around the world. A second area that will dramatically benefit not only animal geography, but the larger human-animal studies community is to continue to map specific practices to make them visible to us on the landscape. We need to know where the CAFOs, the labs, the wildlife sanctuaries, and the pet stores and dog parks are before we can understand how they are affecting animals and humans. Moreover, we need better mapping of all the commodity chains that involve animals as whole beings or their parts to improve our understanding of the economic processes that cause some animals to be worth more (or cost less) than others. The legal structures that shape animals make up an even larger box that needs to be open and mapped. Research into the ethical/political conflicts over the use of animals by humans is a third direction with tremendous potential because these conflicts are the nexus points that alert us to specific practices. Finally, and perhaps the most difficult future direction, is developing the methodologies that will allow us to move closer to the animals themselves as individual, subjective beings that share our planet. As we have seen, several animal geographers have already begun to delineate what this approach might look like, but it has only been applied to very few individuals and species. What, methodologically, does it take to study the subjectivities of species as diverse in both habitat and being as a lemur, a hammerhead shark, a hummingbird, and a tiger?

Obviously working to answer these questions cannot be done in disciplinary (or even subdisciplinary) isolation, and while this book has focused almost exclusively on the animal geography literature, that literature draws from many other fields. Animal geographers of the future will need to build bridges not only with fellow biogeographers (which is already happening), but also with other disciplines like ethology, political science, economics, and conservation biology. The future of animal geography then looks quite promising, and it is well positioned to make itself one of the foundational perspectives within human-animal studies.

PRACTICING AND LIVING ANIMAL GEOGRAPHY

Take a camera out into your daily life and document the geographic analytic categories (historical, economic, cultural landscape, and ethical/political geographies) and concepts (space, scale, place, landscape, and power) that we have focused on. Compile these images and your reflections in an electronic or scrapbook format. This exercise is obviously subjective, and the goal is not to take professional quality photographs, but simply to use your animal geography "eye" to see the cultural animal landscape around you. At this point you should be quite adept at knowing what to look for!

Hopefully the end of this book is not the end of your interest in animal geography and human-animal relationships. If we return to the quote by Henry Beston that opened the book—"Animals are not brethren, they are not underlings; they are other nations, caught with ourselves in the net of life and time, fellow prisoners of the splendor and travail of the earth"—the full weight of it should now be more clear. Whatever our collective and personal experience with nonhumans might be in particular places and at particular times, we cannot deny that our human society has always been deeply and intimately connected to animals. We have not always seen them but they have always been there. Animal geography has given you the tools to "see" animals; their invisibility for you now can only happen by your choice. It is up to you to determine what kind of human-animal earth you want to experience in your time here.

Index

About the Author

Julie Urbanik is an assistant teaching professor in the Department of Geosciences at the University of Missouri–Kansas City, where she teaches courses in human geography and environmental studies. She holds a Ph.D. in geography from Clark University and an M.A. in women's studies from the University of Arizona. Broadly speaking, she is a cultural geographer motivated to explore how issues of identity, place, and technology are (re)configuring human relationships with the natural world. She is the cofounder of the Animal Geography Specialty Group of the Association of American Geographers.